wheat gluten protein analysis

Edited by

Peter R. Shewry

Long Ashton Research Station
Bristol, United Kingdom

and

George L. Lookhart

USDA-ARS
Grain Marketing and Production Research Center
Manhattan, Kansas, U.S.A.

AACC
INTERNATIONAL

Advancing grain science

T0295247

Cover: gluten picture courtesy of the Long Ashton Research Station, Bristol, United Kingdom

This book has been reproduced directly from computer-generated copy submitted in final form to AACC International by the authors. No editing or proofreading has been done by the Association.

Reference in this volume to a company or product name is intended for explicit description only and does not imply approval or recommendation of the product to the exclusion of others that may be suitable.

Library of Congress Catalog Card Number: 2002114116
International Standard Book Number: 1-891127-32-2

Printed in the United States of America on acid-free paper

AACC International, Inc.
3340 Pilot Knob Road
St. Paul, Minnesota 55121 U.S.A.

Preface

The analysis of wheat gluten proteins has a long and distinguished history, extending over a period exceeding 250 years. This reflects the status of wheat as one of the three major cereal crops that dominate world agriculture (the others being maize and rice) and in particular, its preeminent position as raw material for food processing. Its wide use in the food industry depends mainly on the properties of the gluten proteins, which confer viscoelastic properties to doughs, allowing the production of a range of foods including bread, other baked goods, pasta and noodles.

It is not surprising, therefore, that a massive literature has developed on the analysis and properties of wheat gluten proteins, with a particular emphasis on explaining the differences in the functional (i.e. processing) properties that occur between grain samples in terms of variation in the amounts, properties and interactions of individual gluten protein components. Because gluten proteins have unusual properties, being soluble in alcohol/water mixtures and often present as high molecular mass polymers, much of this literature is published in specialist journals and can be confusing to the non-expert.

The present volume therefore, aims to provide standard protocols for the extraction and analysis of wheat gluten proteins based on methods that have been tried and tested in the authors' laboratories. Extensive practical details and tips are provided, as well as suggestions for modifications and examples of applications.

We hope that it will prove of value to established cereal chemists as well as those just entering the exciting field of wheat protein chemistry.

Finally, we are greatly indebted to Mrs. Valerie Topps (Long Ashton Research Station) for her painstaking work in ensuring the quality of the final text.

Peter R. Shewry
George L. Lookhart

Contents

Chapter 1

Wheat Gluten Proteins

Peter R. Shewry
Long Ashton Research Station, Department of Agricultural Sciences,
University of Bristol, Long Ashton, Bristol BS41 9AF, UK

Introduction

Wheat gluten proteins are of immense importance in the food industry as their properties underpin the processing of wheat flour to produce bread, other baked goods, pasta and noodles and a range of other foods. These properties have fascinated scientists for many years, starting with the first description of the isolation of gluten by Jacopo Beccari (Professor of Chemistry at the University of Bologna) in 1745. This has been followed by an immense volume of work, in academic, government and industrial laboratories.

These studies have revealed that gluten is a complex mixture of proteins which vary in their proportions, structures and properties both within and between genotypes. This high level of polymorphism means that characterizing wheat gluten proteins using classical biochemical approaches can be a formidable task. Furthermore, a highly complex and specialised nomenclature has developed which can be impenetrable to the non-specialist.

This chapter therefore aims to provide a concise account of the classification, structures and properties of wheat gluten proteins, in relation to their biological and functional properties.

Gliadins and Glutenins

Beccari's work showed that wheat flour could be separated into two fractions, "amylo" (starch) which was soluble in water and had similar properties to sugars, and "glutinin" (gluten) which was sticky, insoluble in water and resembled substances of animal origin (i.e. proteins). Parmentier (1773) subsequently showed that gluten was largely soluble in vinegar and particularly soluble in spirits of wine while Einhof (1805) showed that gluten was partially soluble in alcohol/water mixtures, with similar fractions also being present in grains of barley and rye. Taddei (1819) also separated wheat glutenin into two fractions that were soluble or insoluble in alcohol, the latter being called "zymom".

These studies provided the basis for the systematic analyses of T.B. Osborne, which put the study of wheat proteins, and indeed the study of plant proteins in general, on a sound scientific basis. Osborne studied seed proteins from some 32 different species (see Osborne 1924) and concluded that the major types of proteins comprised four groups which could be defined on the basis of their sequential extraction in a series of solvents. The first two "Osborne fractions" are extracted in water (albumins) and dilute salt solutions (globulins) and comprise mainly metabolic proteins when extracted from wheat flour, although some minor storage proteins may also be present. In contrast, the third and fourth fractions comprise mainly gluten proteins. The third fraction was extracted with alcohol/water mixtures (typically 60/70% ethanol) and was characterized by high contents of proline and amide nitrogen (now known to be derived from glutamine), hence the name prolamin was coined. This fraction appears to be restricted to the seeds of cereals and other grasses with trivial names being used for fractions from different species: gliadins in wheat, hordeins in barley, secalins in rye etc.

The final fraction was extracted in dilute acid or alkali and was termed glutelin. In wheat this fraction is called "glutenin" but specific names are not used for fractions from other cereals.

In wheat the gliadin and glutenin fractions essentially comprise gluten, being present in approximately equal amounts.

The gliadin/glutenin classification has become widely used for wheat, with the fractionation being applied either to flour as described by Osborne or to gluten. There is no doubt that an important factor determining the durability of the classification has been the differences in the biophysical and functional properties of the two fractions, with elasticity being associated with the glutenins and extensibility with the gliadins (and possibly also small glutenin polymers).

Gliadins and Glutenins are Related but Differ in their Aggregation Behaviour

Early studies of glutenins were limited by their low solubility, except in dilute acid and alkali which can result in some hydrolysis. However, the development in the 1960s of a more effective chaotropic solvent based on urea, a detergent (cetyltrimethylammonium bromide) and dilute acetic acid allowed over 95% of the total gluten proteins to be dissolved and separated by gel filtration chromatography. Such studies demonstrated that the gliadins comprised mainly monomeric proteins (with masses now known to range from about 30,000 to 50,000) while the glutenins were polymeric with masses estimated to exceed 1×10^6 (Meredith and Wren, 1966; Khan and Bushuk, 1979; Field et al 1983). Reduction of disulphide bonds resulted in conversion of the glutenin polymers to monomers, implying that they are stabilized by inter-chain disulphide bonds. Early comparisons of gliadins and glutenin monomers by electrophoresis (Woychik *et al* 1964; Elton and Ewart, 1966; Bietz and Wall, 1972, 1973) and peptide mapping (Ewart, 1966; Bietz and Rothfus, 1970) indicated that the two fractions contained components with related structures and this has since been confirmed by more recent studies (discussed below).

Separation and Classification of Gliadins and Glutenin Subunits

The standard classification used for gliadin subunits is based on their electrophoretic mobility at low pH. The separation was initially carried out by "free boundary" electrophoresis (Jones *et al* 1959) followed by the use of starch gel as a matrix (Woychik *et al* 1961). However, it is now more usual to use polyacrylamide gels with buffers based on aluminium and/or sodium acetate and lactic acid (see, for example, Lafiandra and Kasarda, 1985; Clements, 1987). Although these gel systems are more difficult to apply than the more widely used SDS-PAGE methods (e.g. Laemmli, 1970) they remain widely used for gliadins because they separate the fraction into discrete groups which are valid in terms of their structural and genetic relationships.

Thus, low pH electrophoresis of gliadins resolves four groups of bands which are called α-gliadins (fastest), β-gliadins, γ-gliadins and ω-gliadins (slowest) (Jones *et al* 1959; Woychik *et al* 1961) (Fig. 1), with individual bands within the groups being designated by numbers (e.g. Sexson *et al* 1978) or merely as "fast" or "slow" (e.g. Ewart, 1977, 1983). More recent studies have demonstrated that the α- and β-gliadins are closely related in terms of their amino acid sequences and hence both groups are often referred to as α-type gliadins. However, it has also been shown that some gliadins with α-type sequences are also present in the γ-gliadin region of the gel (Kasarda *et al* 1987).

Early electrophoretic analyses of reduced glutenin subunits by SDS-PAGE showed multiple components and comparison with standard proteins of known molecular mass indicated that these ranged in mass from below 30,000 to about 130,000 (Huebner and Wall 1974; 1975; Bietz and Wall, 1975; Khan and Bushuk, 1977). However, the first definitive study of glutenin subunit composition was reported by Payne and Corfield (1979) who prepared total unreduced glutenin by gel filtration under denaturing conditions before separating the reduced subunits by SDS-PAGE. They defined three groups of subunits, A, B and C, with the B and C groups of subunits having similar masses to the α/β/γ-gliadins and the A subunits having apparent masses above 100,000. Consequently, the A subunits are usually called the high molecular weight (HMW) subunits of glutenin and the B+C groups the low molecular weight (LMW) subunits. More recent studies have also identified a fourth but minor group of LMW subunits (D)

3

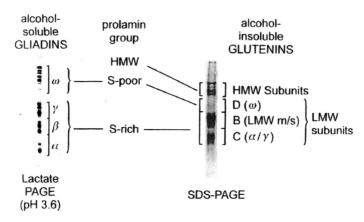

Fig. 1. The classification and nomenclature of wheat gluten proteins separated by SDS-PAGE and electrophoresis at low pH. The D group of LMW subunits are only minor components and are not clearly resolved in the separation shown. (From Shewry *et al* 1999 with permission of Kluwer Academic Publishers).

which migrate slightly slower than C subunits on SDS-PAGE (Jackson *et al* 1985; Payne *et al* 1988) (Fig. 1).

Furthermore, although gliadins are classically considered to be monomeric proteins, gel filtration of alcohol-soluble fractions demonstrates the presence of some polymeric components which can be shown by SDS-PAGE under reducing conditions to comprise bands with similar mobilities to the LMW subunits of glutenin (Bietz and Wall, 1973; 1980; Shewry *et al* 1983). These alcohol-soluble glutenin polymers have been called high molecular weight gliadin or aggregated gliadin although, it is now clear that they represent the low end of the glutenin polymer molecular weight distribution.

Molecular Relationships of Wheat Gluten Proteins

The classification of wheat gluten proteins discussed above is based on differences in their solubility (determined largely by their presence as monomers or as subunits of disulphide-stabilized polymers) and their electrophoretic properties, either in the native state (gliadins) or as reduced subunits (glutenins). This is summarised in Fig. 2.

Although further information was provided by comparison of amino acid compositions, definitive information on the precise relationships of the various gluten proteins was not available until the application of protein sequencing in the 1970s (Kasarda *et al* 1974; Bietz *et al* 1977) followed by molecular cloning of gluten protein-related cDNAs and genes during the following decade (Bartels and Thompson, 1983; Thompson *et al* 1983; Forde *et al* 1983; Kasarda *et al* 1984). As a result we now know the

4

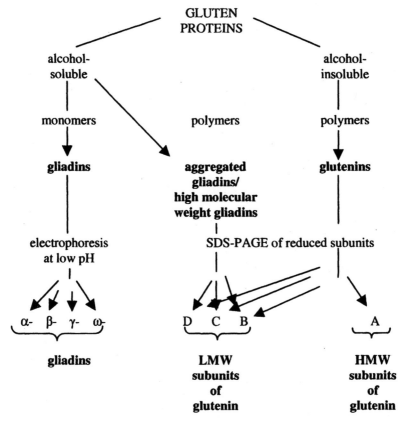

Fig. 2. The "classical" nomenclature for wheat gluten proteins

complete amino acid sequences of a large number of gluten proteins, including components of all the major types.

Based on the information available in the early 1980s Miflin and co-workers proposed a novel classification of wheat gluten proteins and related proteins from barley and rye (Miflin *et al* 1983; Shewry *et al* 1986) which, with subsequent expansion and modification, has become widely accepted as complementary to the classical nomenclature discussed above.

This classification (Fig. 3) recognises three major groups of prolamins:

1. The HMW prolamins correspond to the HMW subunits of wheat glutenin. The widely occurring forms have molecular masses of between 67,000 and 88,000, but calculations based on their mobility on SDS-PAGE may give anomalously high apparent masses of about 82-115,000 (Payne *et al* 1981). Cultivars of bread wheat contain either 3, 4 or 5 individual subunits which are classified into lower M_r y-type and higher M_r x-type subunits (Payne *et al* 1981).

5

Amino acid analysis shows that the HMW subunits are rich in glycine ($\cong$17.20 mol%), glutamine ($\cong$32-36 mol%) and proline ($\cong$10-13 mol%) and comparisons of amino acid sequences shows that these amino acids derive from repetitive sequences which form the central parts of the proteins (Fig. 4). The x-type and y-type subunits differ in that the former contains hexa-, nona- and tripeptide motifs while only hexa- and nonapeptides are present in the y-type subunits. Between four and seven cysteines are present, with three (x-type) or five (y-type) being present in the unique N-terminal domain and one in the C-terminal domain.

2. The S-rich prolamins have M_r by SDS-PAGE of about 30-45,000 and are characterized by high levels of cysteine ($\cong$2-3 mol%), proline ($\cong$15-18 mol%) and glutamine ($\cong$37-40 mol%). Three major families can be recognised based on their amino acid sequences, corresponding to α-type gliadins, γ-type gliadins and B-type LMW subunits of glutenin. All have similar domain structures with short unique N-terminal sequences followed by a repetitive domain and a longer non-repetitive domain (Fig. 4). Similarly, in all cases the repetitive domain is rich in proline and glutamine and contains repeats based on one or two short motifs, although the consensus sequences of these motifs vary between the groups (Fig. 4).

All γ-type gliadins contain eight conserved cysteine residues which are located in the C-terminal domain and form four intra-chain disulphide bonds (Fig. 4). Six of these cysteines, and the three disulphide bonds that they form, are also present in the α-type gliadins (cysteines 1, 4, 5, 6, 7, 8) and the B-type LMW subunits (cysteines 1, 2, 3, 4, 5 and 7) (Fig. 4). In addition, the latter contain one or more additional cysteines present in the N-terminal sequence and/or C-terminal domain, which may be involved in the formation of inter-chain disulphide bonds (see Kohler *et al* 1991, 1993, 1994; Shewry and Tatham, 1997).

Despite their similar domain structures, the α-type gliadins, γ-type gliadins and B-type LMW subunits form discrete groups which differ in the sequences of their N-terminal regions and C-terminal domains, in their consensus repeat motifs and in their disulphide bond structures. In addition, the B-type LMW subunits have been divided into two major sub-families called LMWm and LMWs based on their N-terminal amino acid sequences, which are Met.Glu.Thr and Ser.His.Ile., respectively (Lew *et al* 1992).

Whereas the B-type LMW subunits form a distinct group of S-rich prolamins, this distinction does not apply to the C-type (Payne and Corfield, 1979) which appear to comprise a mixture of α-type and γ-type gliadins which are able to form intra-chain disulphide bonds due to the presence of additional unpaired cysteine residues (Lew *et al* 1992; Shewry and Tatham, 1997; Anderson *et al* 2001). These

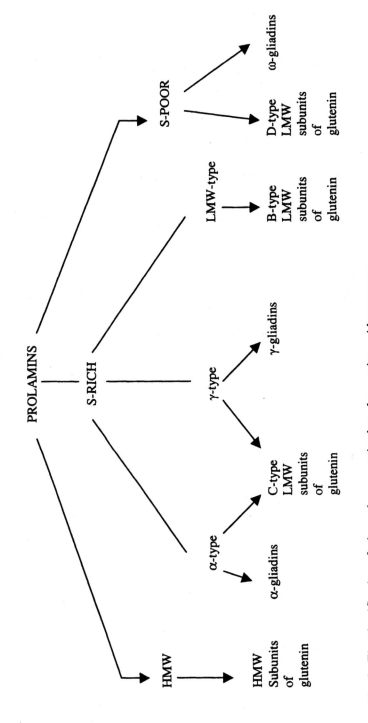

Fig. 3. The classification of wheat gluten proteins based on amino acid sequences.

7

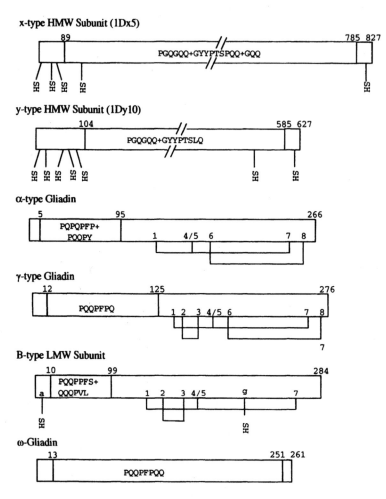

Fig. 4. Summary of the domain structures of "typical" wheat gluten proteins. The consensus amino acid sequences of repeat motifs are shown using standard single letter abbreviations: G, glycine; F, phenylalanine; L, leucine; P, proline; Q, glutamine; S, serine; T, threonine; V, valine; Y, tyrosine. In the gliadins and LMW subunit 1-8 indicate the positions of conserved cysteine residues that form intra-chain disulphide bonds. a and g indicate cysteine residues that may form inter-chain disulphide bonds.

components are, therefore, sometimes called "bound" α-type and γ-gliadins as opposed to the free forms present in the gliadin fraction.

3.   The S-poor prolamins

The ω-gliadins differ from all other gluten proteins in consisting almost solely of repetitive sequences, with only short *N*- and *C*-terminal sequences. They also lack cysteine residues and hence are unable to form either intra- or inter-chain disulphide bonds. Only two full length sequences of ω-gliadins are available (Hsia and Anderson, 2001), with the example shown in Fig. 4 being encoded by the D genome of bread wheat and consisting of short *N*- and *C*-terminal sequences (13 and 10 residues respectively) flanking repeated sequences based on a consensus octapeptide motif (PQQPFPQQ). The latter is related to the consensus motifs in the S-rich prolamins, particularly the γ-gliadins in which the consensus motif differs by the absence of a single glutamine residue (consensus PQQPFPQ). Comparisons of amino acid compositions and *N*-terminal amino acid sequences indicate that most ω-gliadins have similar structures to the example shown in Fig. 4 (Kasarda *et al* 1983; Tatham and Shewry, 1995), except those encoded by the B genome which have a different consensus repeat motif ((Q)(Q)QQXP) (DuPont *et al* 2000).

As with the γ-type and γ-type gliadins, variant forms of ω-gliadins also occur in which the presence of single cysteine residues allows their incorporation into glutenin polymers. These "bound" ω-gliadins correspond to the minor D-type LMW subunits of glutenin that are discussed above (Masci *et al* 1993, 1999).

The Polymorphism and Genetics of Wheat Gluten Proteins

It is clear from the discussion above that wheat gluten is a highly complex mixture of proteins, and this is illustrated by the one and two two-dimensional gel separations shown in Figs. 5 and 6, respectively.

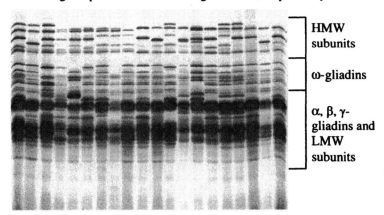

HMW subunits

ω-gliadins

α, β, γ-gliadins and LMW subunits

Fig. 5. SDS-PAGE of seed proteins from twenty cultivars of bread wheat showing polymorphism in the patterns of gluten proteins. (Figure generously provided by Dr. J.M. Field).

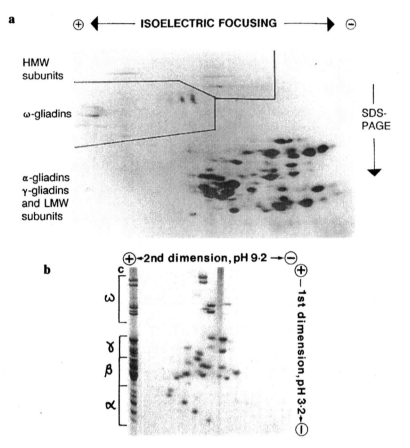

a   ⊕ ◄──── ISOELECTRIC FOCUSING ────► ⊖

HMW
subunits

ω-gliadins   SDS-
PAGE

α-gliadins
γ-gliadins
and LMW
subunits

b   ⊕◄2nd dimension, pH 9·2 ──►⊖

c

ω

γ

β

α

1st dimension, pH 3·2◄──⊕

Fig. 6. 2-D analysis of gluten proteins from bread wheat cv. Chinese
Spring. a, total reduced proteins separated by isoelectric focusing followed
by SDS-PAGE; b, gliadins separated by electrophoresis at pH 3.2 followed
by at pH 9.2 (part b generously provided by Dr. Domenico Lafiandra).

This polymorphism arises in four ways.
1. Polyploidy
    The two major species of wheat cultivated today are polyploids, with
    pasta or durum wheat being a form of the tetraploid species *Triticum
    turgidum* (genome constitution AB) and bread wheat being hexaploid
    (*T. aestivum*) (genome constitution ABD). These cultivated wheats
    arose from the hybridization of related wild diploid species of *Triticum*
    and *Aegilops* with which they share the A, B and D genomes (Feldman,
    1995). Genes encoding gluten proteins are therefore present on all

three genomes resulting in a higher level of complexity in bread wheat than in pasta wheat, with the latter also showing greater complexity than related diploids such as the "primitive" cultivated species *T. monococcum* (einkorn) (genome constitution AA). This is illustrated by the SDS-PAGE comparisons in Fig. 7.

2. Multiple Gene Loci

Analysis of the segregation patterns of gluten proteins in crosses has allowed many of the individual components to be mapped to genetic loci on the A, B and D genomes (Table 1). Thus, the HMW subunits of glutenin are encoded by the *Glu-1* loci on the long arms of the group 1 chromosomes (1A, 1B, 1D) and the α-type gliadins by the *Gli-2* loci on the short arms of the group 6 chromosomes. The γ-type gliadins, ω-gliadins and B-type LMW subunits are encoded by a series of related loci (*Gli-1*, *Gli-3*, *Glu-3*) on the short arms of the group 1 chromosomes (see reviews by Payne, 1987; Shepherd, 1996; Shewry *et al* 1999).

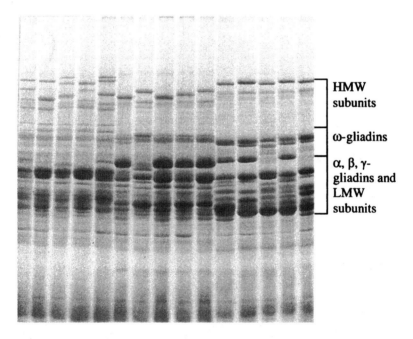

Fig. 7. SDS-PAGE of total seed protein fractions from five genotypes each of hexaploid bread wheat (*T. aestivum*), tetraploid pasta wheat (*T. turgidum* var *durum*) and diploid einkorn (*T. monococcum* var *monococcum*) (Figure generously provided by Dr. R. Fido).

Table 1 Designation and chromosomal location of loci encoding gluten protein in hexaploid bread wheat (*T. aestivum*).

Locus	Location	Proteins encoded
Gli-A1 *Gli-B1* *Gli-D1*	1A, 1B, 1D short arms	γ-gliadins ω-gliadins
Gli-A3 *Gli-B3*	1A, 1B short arms	ω-gliadins D-type LMW subunits
Glu-A3, Glu-B3 *Glu-D3*	1A, 1B, 1D short arms	B-type LMW subunits
Gli-A2, Gli-B2 *Gli-D2*	6A, 6B, 6D short arms	α-type gliadins
Glu-A1 *Glu-B1* *Glu-D1*	1A, 1B, 1D long arms	HMW subunits

3. Multigene Families

Molecular analyses have demonstrated that all of the gluten protein loci consist of multigene families, the simplest being the *Gli-1* loci which each comprises two genes encoding one x-type and one y-type subunit. The total numbers of copies of genes for other gluten proteins are not precisely known with estimates varying widely. For example, Harberd *et al* (1985) reported 25-35 copies of α-gliadin genes per haploid genome compared with over 100 copies by Okita *et al* (1985) and up to 150 copies by Anderson *et al* (1997). Similarly, Sabelli and Shewry (1991) estimated the presence of between 17 and 39 γ-gliadin genes, 15-18 ω-gliadin genes and 22-35 B-type LMW subunit genes in bread wheat cv. Chinese Spring, the variation depending on the restriction enzyme used for molecular analysis.

However, it is clear that not all of the gluten protein genes in wheat are expressed. The best characterized example of gene silencing is at the *Gli-1* loci, where variation in the number of expressed subunits from three to five is associated with silencing of the 1Ax, 1By and 1Ay genes. Similarly, Anderson and Greene (1997) compared the sequences of 27 known cDNA and genomic clones for α-type gliadins and concluded that about half of the latter contained "in frame" stop codons and were presumably pseudogenes. Hsia and Anderson (2001) isolated two ω-gliadin genes, one of which was a pseudogene, while Rafalski (1986) isolated a γ-gliadin pseudogene. In contrast, Cassidy *et al* (1998) were unable to identify pseudogenes in the seventeen sequences that were then available for B-type LMW subunits.

It is probable, therefore, that the extent of gene silencing varies between different gluten protein gene families and loci.

4. Allelic Variation

    There is a high level of allelic variation in the gluten protein patterns of different genotypes of wheat, which includes variation in the amounts, absence or presence and properties (including electrophoretic mobility) of components. This is illustrated by the comparison of SDS-PAGE patterns of gluten proteins from a range of bread wheat cultivars in Fig. 5.

Implications for Gluten Protein Analysis

The unusual solubility properties of wheat gluten proteins and the presence of highly complex mixtures of components, many of which have similar molecular masses and biochemical properties, pose formidable problems for the cereal chemist. This in turn has resulted in a continued interest in the development of methods, with modifications to existing methods and even entirely new methods being regularly reported in specialist journals. Consequently, it is difficult for scientists entering the field of wheat gluten chemistry to decide on the most appropriate approach. The following chapters in this book aim to eliminate, or at least alleviate, this problem by providing detailed protocols for methods that are widely accepted to "work". These are supported by practical tips, recommended modifications for specific purposes, and examples of applications. It is intended that the result will be of interest and value to established cereal chemists as well as new entrants to the field.

Acknowledgement

Long Ashton Research Station receives grant-aided support from the Biotechnology and Biological Sciences Research Council of the United Kingdom.

References

Anderson, O.D. and Greene, F.C. 1997. The α-gliadin gene family. II. DNA and protein sequence variation, subfamily structure, and origins of pseudogenes. Theor. Appl. Genet. 95:59-65.

Anderson, O.D., Hsia, C.D. and Torres, V. 2001. The wheat γ-gliadin genes: characterization of ten new sequences and further understanding of γ-gliadin gene family structure. Theor. Appl. Genet. 103:323-330.

Anderson, O.D., Litts, J.C. and Greene, F.C. 1997. The α-gliadin gene family. I. Characterization of ten new wheat α-gliadin genomic clones, evidence for limited sequence conservation of flanking DNA, and southern analysis of the gene family. Theor. Appl. Genet. 95:50-58.

Bartels, D. and Thompson, R.D. 1983. The characterization of cDNA clones coding for wheat storage proteins. Nucleic Acids Res. 11:2961-2977.

Beccari 1745. De Frumento. De Bononiensi Scientiarum et Artium Instituto atque Academia Commentarii, II., Part I., p. 122-127.

Bietz, J.A. and Rothfus, J.A. 1970. Comparison of peptides from wheat gliadin and glutenin. Cereal Chem. 47:381

Bietz, J.A. and Wall, J.S. 1972. Wheat gluten subunits: molecular weights determined by sodium dodecyl sulfate-polyacrylamide gel electrophoresis. Cereal Chem. 49:416-429.

Bietz, J.A. and Wall, J.S. 1973. Isolation and characterization of gliadin-like subunits from glutenin. Cereal Chem. 50:537-547.

Bietz, J.A. and Wall, J.S. 1975. The effects of various extractants on the subunit composition and associations of wheat glutenin. Cereal Chem. 52:145-155.

Bietz, J.A., Huebner, F.R., Sanderson, J.E. and Wall, J.S. 1977. Wheat gliadin homology revealed through N-terminal amino acid sequence analysis. Cereal Chem. 54:1070-1983.

Cassidy, B.G., Dvorak, J. and Anderson, O.D. 1998. The wheat low-molecular-weight glutenin genes: characterization of six new genes and progress in understanding gene family structure. Theor. Appl. Genet. 96:743-750.

Clements, R.L. 1987. A study of gliadins of soft wheats from the Eastern United States using a modified polyacrylamide gel electrophoresis procedure. Cereal Chem. 64:442-448.

DuPont, F.M., Vensel, W.H., Chan, R. and Kasarda, D.D. 2000. Characterization of the 1B-type ω-gliadins from *Triticum aestivum* cultivar Butte. Cereal Chem. 77:607-614.

Einhof, H. 1805. Chemische analyse de roggens (Secale cereale). Neues allgem. J. d. Chem. 5:131-153

Elton, G.A.H. and Ewart, J.A.D. 1966. Gluteinins and gliadins: electrophoresis studies. J. Sci. Food Agric. 17:34-38.

Ewart, J.A.D. 1966. Fingerprinting of gluteinin and gliadin. J. Sci. Food Agric. 17:30-33.

Ewart, J.A.D. 1977. A fast-moving β-gliadin from Cappelle-Desprez. J. Sci. Food Agric. 28:1080-1083.

Ewart, J.A.D. 1983. Slow triplet β-gliadin from Cappelle-Desprez. J. Sci. Food Agric. 34:653-656.

Feldman, M., Lupton, F.G.H. and Miller, T.E. 1995. Wheats *Triticum* spp. (Gramineae-Triticinae). Pages 184-192 in: Evolution of Crop Plants. J. Smartt and N. W. Simmonds, eds. Longman Scientific & Technical, Harlow, U.K.

Field, J.M., Shewry, P.R., and Miflin, B.J. 1983 Solubilization and characterization of wheat gluten proteins; correlations between the amount of aggregated proteins and baking quality. *J. Sci. Food Agric.* 34:370-377.

Forde, J., Forde, B.G., Fry, R.P., Kreis, M., shewry, P.R. and Miflin, B.J. 1983. Identification of barley and wheat cDNA clones related to the high M_r polypeptides of wheat glutenin. Febs Lett. 162:360-366.

Harberd, N.P., Bartels, D. and Thompson, R.D. 1985. Analysis of the gliadin multigene loci in bread wheat using nullisomic-tetrasomic lines. *Mol. Gen. Genet.* 198:234-242.

Hsia, C.C. and Anderson, O.D. 2001. Isolation and characterization of wheat ω-gliadin genes. *Theor. Appl. Genet.* 103:37-44.

Huebner, F.R. and Wall, J.S. 1974. Wheat glutenin subunits. I. Preparative separation by gel-filtration and ion-exchange chromatography. Cereal Chem. 51:228-240.

Huebner, F.R. and Wall, J.S. 1975. Glutenin from wheat varieties varying in baking quality. Cereal Chem. 53:258-269.

Jackson, E.A., Holt, L.M. and Payne, P.I. 1985. *Glu-B2*, a storage protein locus controlling the D group of LMW glutenin subunits in bread wheat (*Triticum aestivum*). Genet. Res. Camb. 46:11-17.

Jones, R.W., Taylor, N.W. and Senti, F.R. 1959. Electrophoresis and Fractionation of Wheat Gluten. Arch. Biochem. Biophys.. 84:363-376.

Kasarda, D.D., Da Roza, D.A. and Ohms, J.I. 1974. α_2-gliadin. Biochim. Biophys. Acta 351:290-204.

Kasarda, D.D., Autran, J-C., Lew, E.J-L., Nimmo, C.C. and Shewry, P.R. 1983. *N*-terminal amino acid sequences of ω-gliadins and ω-secalins: implications for the evolution of prolamin genes. Biochim. Biophys. Acta 747:138-150.

Kasarda, D.D., Okita, T.W., Bernardin, J.E., Baecker, PA., Nimmo, C.C., Lew, E.J.-L., Dietler, M.D. and Greene, F. 1984. Nucleic acid (cDNA) and amino acid sequences of α-type gliadins from wheat (*Triticum aestivum*). Proc Natl. Acad. Sci. U.S.A. 81:4712-4716.

Kasarda, D.D., Adalsteins, A.E. and Laird, N.F. 1987. Gamma-gliadins with α-type structure coded on chromosome 6B of the wheat (*Triticum aestivum* L.) cultivar 'Chinese Spring,' pages 20-29, in Proceedings of the Third International Workshop on Gluten Proteins, Lasztity, R. and Békés, F. (eds.). World Scientific Publishing, Singapore.

Khan, K. and Bushuk, W. 1977. Studies of glutenin. IX. Subunit composition by sodium dodecyl sulfate-polyacrylamide gel electrophoresis at pH 7.3 and 8.9. Cereal Chem. 54:588-596.

Khan, K. and Bushuk, W. 1979. Studies of glutenin. XIII. Gel filtration, isoelectric focusing, and amino acid composition studies. Cereal Chem. 56:505-512.

Köhler, P., Belitz, H.-D. and Wieser, H. 1991. Disulphide bonds in wheat gluten: isolation of a cysteine peptide from glutenin. Z. Lebensm. Unters. Forsch. 192:234-239.

Köhler, P., Belitz, H.-D. and Wieser, H. 1993. Disulphide bonds in wheat gluten: further cysteine peptides from high molecular weight (HMW) and low molecular weight (LMW) subunits of glutenin and from γ-gliadins. Z. Lebensm. Unters. Forsch. 196:239-247.

Köhler, P., Keck, B., Müller, S. and Wieser, H. 1994. Disulphide bonds in wheat gluten. Pages 45-54 in Wheat Kernal proteins, Molecular and Functional Aspects. Proc. Int. Meeting S. Martino at Cimino, Viterbo, Italy, 28-30 Sept .

Lafiandra, D. and Kasarda, D.D. 1985. 1- and 2-dimensional (2 pH) system of polyacrylamide gel electrophoresis carried out in a single cell: separation of wheat proteins. Cereal Chem. 62:314-319.

Leammli, U.K. 1970. Cleavage of structural proteins during the assembly of the head of bacteriophage T4. Nature 227:680-685.

Lew, E.J.-L., Kuzmicky, D.D. and Kasarda, D.D. 1992. Characterization of low molecular weight glutenin subunits by reversed-phase high performance liquid chromatography, sodium dodecyl sulfate-poly-acrylamide gel electrophoresis, and *N*-terminal amino acid sequencing. Cereal Chem. 69:508-515.

Masci, S., Lafiandra, D., Porceddu, E., Lew, E.J.-L., Tao, H. P. and Kasarda, D.D. 1993. D-glutenin subunits: *N*-terminal sequences and evidence for the presence of cysteine. Cereal Chem. 70:581-585.

Masci, S., Egorov, T.A., Ronchi, C., Kuzmicky, D.D., Kasarda, D.D. and Lafiandra, D. 1999. Evidence for the presence of only one cysteine residue in the D-type low molecular weight subunits of wheat subunits of wheat glutenin. J. Cereal Sci. 29:17-25.

Meredith, O.B. and Wren, J.J. 1966. Determination of molecular-weight distribution in wheat-flour proteins by extraction and gel filtration in a dissociating medium. Cereal Chem. 43:169

Miflin, B.J. Field, J.M., Shewry, P.R. 1983. Cereal storage proteins and their effects on technological properties. Pages 255-319 in: Seed Proteins. J. Daussant, J. Mosse and J. Vaughan, eds. Academic Press: London.

Okita, T.W., Cheesbrough, V. and Reeves, C.D. 1985. Evolution and heterogeneity of the α-/β-type and γ-type gliadin DNA sequences. J. Biol. Chem. 260:8203-8213.

Osborne, T.B. 1924. *The Vegetable Proteins*; 2nd edn. Longmans Green and Co.: London.

Parmentier, A.A. 1773. Examin chimique des pommes de tere, dans lequel on traite des parties constituantes du blé. Didot le jeaune, Paris.

Payne, P.I. 1987. Genetics of wheat storage proteins and the effect of allelic variation on bread-making quality. Ann. Review Plant Physiol. 38:141-153.

Payne, P.I. and Corfield, K.G. 1979. Subunit composition of whet glutenin proteins, isolated by gel filtration in a dissociating medium. Planta 145:83-88.

Payne, P.I., Holt, L.M. and Law, C.N. 1981. Structural and genetical studies on the high-molecular-weight subunits of wheat glutenin. Theor. appl. Genet. 60:229-236.

Payne, P.I., Holt, L.M. and Lister, P.G. 1988. *Gli-A3* and *Gli-B3*, two newly designated loci coding for omega-type gliadins and D subunits of glutenin. Pages 999-1002 in Proc.7th International Wheat Genetics Symposium, 13-19 July. Institute of Plant Science Research, Cambridge.

Rafalski, J.A. 1986. Structure of wheat gamma-gliadin genes. Gene 43:221-229.

Sabelli, P. and Shewry, P.R. 1991. Characterization and organisation of gene families at the *Gli-1* loci of bread and durum wheats by restriction fragment analysis. Theor. Appl. Genet. 83:209-216.

Sexson, K.R., Wu, Y.V., Huebner, F.R. and Wall, J.S. 1978. Molecular weights of wheat γ₂-. β6-, α7-, α8- and α9-gliadins. Biochim. Biophys. Acta, 532:279-285.

Shepherd, K.W. 1996. Gluten genetics - a perspective after 30 years. Pages 8-11 in Gluten '96, Proceedings of the 6th International Gluten Workshop. C.W. Wrigley, ed. Royal Australian Chemical Institute, North Melbourne, Australia.

Shewry, P.R. and Tatham, A.S. 1997. Disulphide bonds in wheat gluten proteins. J. Cereal Sci. 25:207-227.

Shewry, P.R., Miflin, B.J., Lew, E.J-L. and Kasarda, D.D. 1983. The preparation and characterization of an aggregated gliadin fraction from wheat. J. Exp. Bot. 34:1403-1410.

Shewry, P.R., Tatham, A.S. and Halford, N.G. 1999. The prolamins of the Triticeae. Pages 33-78 in Seed Proteins. P.R. Shewry and R. Casey, eds. Kluwer Academic Publishers: Dordrecht.

Shewry, P.R., Tatham, A.S., Forde, J., Kreis, M. and Miflin, B.J. 1986. The classification and nomenclature of wheat gluten proteins: a reassessment. J. Cereal Sci. 4:97-106.

Taddei, G. 1819. Ricerche sul glutine del frumento. Giornale di fisica, chimica, e storia naturale, Brugnatelli, 2:360-361

Tatham, A.S. and Shewry, P.R. 1995. The S-poor prolamins of wheat, barley and rye. J. Cereal Sci. 22:1-16.

Thompson, R.D., Bartels, D., Harberd, N.P. and Flavell, R.B. 1983. Characterization of the multigene family coding for HMW glutenin subunits in wheat using cDNA clones. Theor. Appl. Genet. 67:87-96.

Woychik, J.H., Boundy, A. and Dimler, R.J. 1961. Starch gel electrophoresis of wheat gluten proteins with concentrated urea. Arch. Biochem. Biophys. 94:477-482.

16

Woychik, J.H., Heubner, F.R. and Dimler, R.J. 1964. Reduction and starch-gel electrophoresis of wheat gliadin and glutenin. Archives Biochem. Biophys. 105:151-155.

Chapter 2

Fractionation Techniques

Rob. J. Hamer
Centre for Protein Technology TNO- Wageningen University

Introduction

Wheat flour is a complex mixture of different components. The proteins of wheat are a complex mixture of often closely related polypeptides. It is sometimes necessary to fractionate and/or quantify these proteins. The gluten content of wheat and the ratio between gliadin and glutenin are two examples. More advanced studies often require the isolation of individual protein subtypes. In the scientific literature scores of publications have appeared describing isolation protocols. It is not the intention of this chapter to review all these publications. The chapter is intended for the reader with basic experience in protein purification, to assist in planning the fractionation or isolation of wheat proteins. This leads to a set of rules specific to wheat proteins that can be combined with isolation protocols or even published methods to come to an optimised protocol. The chapter will be structured around two concepts: Analysis and Preparation. Both are related to the purpose of the fractionation. Is the purpose of fractionation to quantify a certain fraction (Analysis) or is the fraction itself to be used for further research on the functionality of the proteins present (Preparation)? The chapter will not deal with specific issues related to preparative fractionation on an industrial scale.

The First Question: Analytical or Preparative?

It stands to reason that there are more conditions to be fulfilled when the fractionation serves the purpose of preparing a protein to be studied further. The issues involved will be discussed in the following section. In this section I will focus on the analytical fractionation of proteins.

Analytical Fractionation

The following examples serve to illustrate the value of analytical fractionation:

- the classification of wheat using Reversed-Phase High Performance Liquid Chromatography (RP-HPLC) or Capillary Electrophoresis (CE)
- the gluten content of wheat flour

- the ratio of gliadin/glutenin
- the content of SDS-insoluble glutenin in flour and dough samples

In all cases the purpose is to quantify a certain fraction of wheat protein. Such quantification could be used to classify a wheat variety or a wheat flour sample. Examples of such techniques include RP-HPLC (Bietz, 1985) and CE (Bean et al 1998) of wheat gliadin or wheat glutenin. Standard rules for analytical measurements should apply in such a case: the method needs to comply with standard rules of quantification: reproducibility, variability, standardization and calibration. Depending on the methods at hand, these criteria are not always easy to meet. For example, the measurement of SDS-insoluble glutenin in a flour sample is dependent on the formation of a (glutenin polymer) gel upon ultracentrifugation (Graveland, 1980). The amount of gel is therefore not only dependent on the amount of SDS-insoluble protein in the sample, but also on the centrifugation conditions. On the other hand, the sample itself can be the origin of much variation. If, for example, the sample contains a high amount of damaged starch, the viscosity of the continuous phase can affect the ultracentrifugation.

Characteristics of the Sample that Influence Subsequent Fractionation Procedures

In cereal science the sample is usually wheat flour (from white to wholemeal), dough or a finished product. In almost all cases, the sample has to be dried and ground before extraction. Drying is a necessary step to ensure the stability of the protein in the sample during storage, but the process of drying itself should not influence the protein. This is easier said than done. Depending on the drying conditions used, significant damage to the protein can occur (Lupano et al 1987). It is for this reason that the history of the sample should be known.

A second issue of considerable importance is the particle size of the sample to be extracted. Particle size is related to the surface area and hence can affect the extraction. As a rule of thumb, in all cases the sample should be milled to pass a 0.5 mm sieve. When milling to a smaller particle size, care must be taken to avoid heat generation during milling, since this can cause damage of the protein to occur with consequences for the subsequent extraction.

A third point of importance is the homogeneity of the sample. In particular, if the sample size used for the extraction is small (< 1 g), care must be taken to obtain a representative sub-sample. In such a case, the stock needs to be thoroughly mixed before the sample is taken. Also replicate sub-samples have to be taken.

Fourth, some material is 'treated', and contains either an oxidative chemical (flour improver) or enzymes. As a general rule, such material must be avoided. If for any reason this is not possible, care must be taken to avoid interference of such components and additional checks on the integrity of the protein are advised.

Finally, a point of particular concern is related to the lipid content of the sample. In many wet fractionation procedures, lipid from the original sample can cause problems during fractionation. For this reason, samples are often defatted prior to extraction. A variety of procedures is found in the literature. Samples are extracted with chloroform, petroleum ether, water saturated butan-1-ol or hexane. Again, care must be taken to use a method that does not affect the fractionation procedure.

General Factors Influencing Extraction

Extraction procedures are based on differences in solubility in the solvent used. Large differences exist between the different wheat protein classes that can be used as a basis for their separation. In general the following classes are distinguished:

- Albumins and globulins
- Gliadins
- Low molecular weight glutenin subunits (LMWGS)
- High molecular weight glutenin subunits (HMWGS)
- LMW glutenin polymers
- HMW glutenin polymers
- Glutenin macropolymer (GMP, also called gel protein)

Table 1 gives a summary of the differences in solubility of these classes:

The first and most important message in this respect is that 'nothing is absolute'. A salt-water extract of flour will contain traces of gliadin and LMW glutenin, an EtOH extract will contain HMW glutenin as well as gliadin. If extraction gives a purity of a specific fraction of 80 % or more, this is excellent. The second remark to be made is related to the extent of extraction. An ethanol extract of flour will not contain all the gliadins, and a salt water extract will not contain all the albumin/globulin proteins. This could be a specific point of concern when an assay is used for quantification. One such example is the commercial test for gluten protein (the so-called 'Skerritt' test). The test is based on the extraction of ω-gliadins. This sub-type of gliadin is the ideal protein in terms of its extraction properties. It has no disulfide bonds and is extremely stable and hence can be extracted with most solvents even from a heated product. The test is suitable to identify the presence of gluten protein, but is limited with regard to quantification. The reason for this lies in the fact that extraction is not quantitative in most cases. Also, the test is not sensitive to all gluten protein components and their peptides.

21

Table 1: Differences in solubility that can be exploited in wheat protein fractionation.

Protein class	Soluble in	Remarks	References
Albumin, Globulin	Water (albumins) Dilute salt solution (albumins and globulins)	pH is important, pH must be at least pI±1	
Gliadin	Water, 60 % (v/v) ethanol	Low solubility in water, solubility is improved at low pH (< 4) and low ionic strength	Bietz and Wall, 1973; Popineau, 1985
LMW Glutenin polymers	Water, SDS solution, 70 % (v/v) ethanol, 50% (v/v) propan-1-ol	Applicable to small oligomers only. Low solubility in water.	Bietz and Wall, 1973; Larre et al 1997; Bean et al 1999
LMWGS, HMWGS	Water, SDS solution, 70 % (v/v) ethanol, 50% (v/v) propan-1-ol	Reducing conditions are required to prevent repolymerisation.	
HMW Glutenin polymers	1.5 % (w/v) SDS	Applies only to HMW glutenin polymers < ca 500 kDa	Graveland, 1980
Glutenin macropolymer	Insoluble in the solvents listed above	Only soluble after reduction or sonication	Graveland, 1980; Singh et al 2000

Potential Pitfalls[1]

In a number of extraction protocols, the sample is pre-treated to improve solubility. With gluten proteins either a detergent is used (e.g. sodium dodecylsulphate, SDS), a reducing agent (e.g. dithiothreitol (Bean and Lookhart, 1998), or 2-mercaptoethanol (Nicolas et al 1997)), or sonication (Singh et al 2000). There are a number of potential pitfalls to consider. With DTT, it is the effective concentration that counts. The reagent is prone to oxidation by exposure to oxygen with time, temperature, and pH of exposure being important. Care must therefore be taken to use a freshly prepared stock solution of DTT and an adequate surplus of reducing agent. Also, reaction time and temperature should be kept constant.

---

[1]See also the paragraph "Characteristics of the sample that influence subsequent fractionation procedures."

With a detergent such SDS (sodium dodecylsulfate) a sufficient excess of SDS to protein should ideally be used. The detergent acts by binding to the protein, but is non-specific and potential 'scavengers' (such as fats and lipids) could lower the effective concentration. It is preferable that the concentrations of SDS that are used are higher than its critical micelle concentration (> ca 1 % (w/v)). Furthermore, potassium ions can interfere with SDS (the potassium dodecylsulphate complex is poorly soluble) and the temperature should not be lower than 10 °C.

Sonication is often used to improve the extraction of very high molecular weight glutenin polymeric proteins (Singh *et al* 2000; Gupta *et al* 1992). With sonication energy is transferred to the sample by a vibrating tip. This energy leads to depolymerisation of the high molecular weight glutenin polymers. During this process, substantial heat is generated. This heat could influence the solubility of the protein and the sample is usually cooled on ice during sonication. Other parameters that are important include time and energy/power setting of the sonication device. If the time or power settings are too high breakdown of the actual protein polypeptide chain will occur.

Final Remarks on Quantitative Extraction Procedures

In the cereal science literature many analytical fractionation procedures have been published. Some examples are given in Table 2. With some exceptions (i.e. RP-HPLC of wheat proteins) few of these methods have been scrutinized in terms of standard requirements for analytical tests. It can be even argued that some of the tests reported can never pass such scrutiny since they are very sensitive to sample composition and therefore cannot be standardized.

Table 2: Analytical procedures used to quantify wheat proteins

Procedure	Tests	reference
Gluten content	ICC standard method No 155	
Glutenin composition	RP-HPLC	Bietz, 1983
Gliadin composition	RP-HPLC	Bietz, 1984; Nicolas *et al* 1998
HMW glutenin content	Sonication, Size Exclusion chromatography	Gupta *et al* 1993
Glutenin macropolymer content	SDS extraction/ ultracentrifugation	Graveland, 1980

Preparative Fractionation

Cereal science has profited greatly from the endeavors of cereal scientists who have purified individual protein fractions and used them in so-called reconstitution studies. In this way information can be obtained on the importance of individual protein fractions. This was already started with the famous work of Osborne, and even today, Osborne fractionation is often used to separate the major fractions of wheat protein. More recently, the work of Békés and co-workers (Tamas *et al* 1998, Békés and Gras 1999) with individual HMW glutenin subunits demonstrates that this approach is still valid today. I will also briefly discuss the use of heterologous expression techniques to produce such fractions. A matter of concern with all these studies is whether the protein studied is in its native form. In other words, 'to what extent have the expression or isolation and purification procedures, led to a change in functional properties?' This aspect is the central issue to be discussed in this section.

The Native Protein

'If only gluten was an enzyme'. With enzymes, the story seems relatively simple. You isolate the protein and if it is still active, you can state that you have isolated the protein in its native form. With gluten, the situation is far more complex. First, most of the gluten proteins are not soluble in water. Second, how can we check the functionality of the protein? Any discussion on this topic runs the risk of becoming philosophical. Nevertheless, an attempt is made with the following statement:

"A gluten protein can be considered native, if similar functional properties can be observed with the isolated form as with the native form in dough."

Functionality Criteria Reflecting Composition

This statement contains some key phrases. First, what functional properties need to be considered in this respect. A number of such properties are well accepted and easily defined. These are:
- isoelectric point
- apparent molecular weight
- solubility
- sequence

Such properties reflect the composition of the protein and give information on the integrity of the protein isolated. The advantage of some of these techniques is that in combination with selective detection techniques (such as specific antibodies) they are also suitable for non-fractionated systems. Thus, such properties can be compared between the isolated and non-isolated proteins. To some extent, the same applies to sequence information. From the DNA sequence an amino acid sequence can be derived and hence a mass estimate. This can be compared to the mass determined for the isolated protein.

Criteria Reflecting Conformation

The following criteria that are related to the structure of the protein are often more difficult to use:

- Differential Scanning Calorimetry (DSC) thermogram
- Circular Dichroism (CD) or Infrared (IR) spectrum
- (lack of) molecular aggregation

A DSC thermogram (Chevallier *et al* 1999) can provide useful information provided the protein exhibits a clear thermal transition. Such a transition should then coincide with a similar transition in the starting material. Unfortunately, gluten does not give a clear thermal transition that can be attributed to protein.

A CD and/or IR spectrum gives important information about the conformation and structure elements of the protein, as discussed in detail in Chapter 8. This immediately requires us to define the native conformation of the protein. Following the statement made earlier, we should take the conformation of the protein as it occurs in dough. It is proposed to take dough as the reference system, and not the wheat kernel or flour, since the protein is in a largely dehydrated stored state in the grain and in the flour. Also, in the whole of cereal science, it is in the dough where the gluten proteins exhibit their unique functionality. So let us make the dough the system of reference. The next question therefore is: "What is the conformation of the gluten protein in dough?" There is only scarce literature to refer to here. There is no consensus on the actual aqueous or lipid-like environment of the gluten proteins in dough or gluten (Marion *et al* 1987; Hargreaves *et al* 1994a). Bekkers *et al* (1999) studied the solubility of HMWGS 1Dx5. Expressed recombinant proteins, representing the N- and C-terminal domains (dA and dC, respectively) had poor solubility. The same holds for the complete subunit, isolated from flour. Removal of the dA and dC domains resulted in a 58 kD peptide containing the complete repetitive B domain. This peptide was soluble in aqueous solution. This makes it likely that parts of the peptide are hydrophilic and readily involved in hydrogen bridges with water molecules or other B domains, whereas other parts are not exposed to water and are perhaps present in a more lipophilic environment. Such a situation is in agreement with NMR studies by Belton *et al* (1999) and spin labeling studies (Hargreaves *et al* 1994b).

There are clear reports in the literature about the poor solubility of both gliadin and glutenin. With gliadin the solubility can be improved by lowering the pH. This, however, is accompanied by a changed conformation (Popineau and Pineau, 1988). On the other hand, the isolation of gliadin in the presence of some acid prevents aggregation and allows the isolation of a functional protein (Weegels et al 1995). With glutenin the reduction of disulfide bonds and subsequent modification of sulfhydryl groups is required to obtain reasonable solubility in low pH, low ionic strength buffers. As a consequence, most functionality is lost and a study of network-forming properties is made impossible. High molecular weight polymers are only poorly soluble in such buffers and usually require an organic solvent or a detergent solution (SDS) to solubilize.

Criteria Related to the Actual Functionality of the Protein

Gluten proteins are characteristically known for their rheological properties, including visco-elasticity and strain hardening. Furthermore, they may, together with lipids and other proteins, render gas-retaining properties to dough. These features of gluten are mainly contributed by the HMW and LMW glutenin protein subunits in the presence of gliadin. It is the class of glutenin proteins that is most difficult to isolate in a way that allows its characterization. On the other hand, it is this class of proteins that may hold an important key to understanding functionality. The difficulties with glutenin proteins require another approach to functionality. Today it is possible to isolate individual glutenin proteins and/or produce them in several expression systems (Tamas et al 1998). However, there is no definition of the native conformation of such protein. The only system we can then refer to is the actual dough, or systems which exhibit a functionality strongly related to that of dough. Two such systems are described below.

The Reconstitution System

Although many authors have carried out reconstitution studies, Wrigley et al (1999) provided an early insight by showing that just the addition of a purified HMW glutenin subunit to dough was not sufficient to have an effect on functionality. It is generally accepted that such a subunit is an integral part of the protein network in dough. Hence, they devised a protocol aimed at incorporating the subunit into a background network structure. This is discussed in detail in Chapter 9. Although the relevance of such a system remains to be proven, a number of functionally relevant features of glutenin subunits could be determined.

The Glutenin Macropolymer

In 1980 Graveland *et al* (1980) reported a fractionation system based on the solubility of almost all gluten protein in SDS. The fraction that remained insoluble could be harvested as a gel by centrifugation. Graveland named this fraction 'gel protein'. Later, Weegels *et al* (1997) renamed it "Glutenin Macropolymer (GMP)" to better reflect the fact that this fraction consisted of highly polymerized HMW and LMW glutenin subunits. Furthermore, GMP is relevant to functionality since its quantity, and also its rheological properties closely reflect dough properties (Weegels *et al* 1996). The GMP fraction is well defined. As a disadvantage, it can only be used to study glutenin network functionality. Also, the use of SDS limits the flexibility of the system to study effects of enzymes and other non-covalent interactions. On the other hand, recent studies demonstrate the value of this system to explore the intricate background of glutenin protein functionality (Don *et al* 2002).

Summary of Functionality Tests

Table 3 summarizes the various tests that can be used to check the functionality of the protein purified.

Table 3 does not claim to be exhaustive. The general message from the table is clear: solubility of the material too a large extent determines our ability to study functional properties, even at a molecular level. At a macroscopic level, only reconstitution studies are relevant, since they allow for example large deformation rheological testing of gluten proteins. The drawback of such studies is the uncertainty of a successful re-incorporation

Table 3: Suitability of approaches to study the functionality of isolated proteins.

Protein class	Functionality tests			
	Related to composition: IEF, M_r sequence	Related to conformation: DSC, CD, IR, NMR	Actual function	
			Reconstitution studies	GMP rheology
Albumins, Globulins	+++	+++	++	-
Gliadins	++	±	++	-
LMW Glutenin subunits	±	-	±	++
HMW Glutenin subunits	±	-	±	++

+++: no problem; ++: can be done but attention should be paid to solubility;
±: possible, but not always reliable; -: not feasible or relevant.

of glutenin subunits into the gluten protein network. Physical characterisation of GMP gives information on the glutenin network on a mesoscopic scale. Both the quantity and rheological parameters of GMP correlate with macroscopic properties and are hence relevant. Nevertheless, the discussion on what is native and what is not is a difficult one. Wheat gluten proteins still present a true challenge in defining realistic criteria. Without such criteria, any purification procedure is doomed to failure.

Purification: a Conclusion

It may seem awkward to close a chapter on protein purification with only a short section on the purification of the various protein classes mentioned throughout this chapter. Today, however, the challenge is not to purify individual proteins. The difficulty lies in how to obtain the proteins in such a condition that they can be fractionated and still maintain their relevance to the aim of the researcher. For this reason, considerable attention was spent on discussing the preparation of the material prior to fractionation, on the differential solubility of gluten protein classes and on functionality criteria. Numerous protocols can be found in the literature and adapted using the information in this chapter. Examples of some preparative procedures are given in Table 4.

Research on gluten protein functionality is ongoing; the emphasis is shifting towards higher molecular weight gluten protein fractions and their mesoscopic behaviour. It is expected that this will not only give exciting new results but also new insights that can be used to solve functionality issues and improved purification protocols.

Table 4: Examples of preparative procedures.

Protein isolated	Principle	References
Gliadin	Hydrophobic interaction chromatography, Fast Protein Liquid Chromatography	Popineau and Pineau, 1985; Larre *et al* 1991; Weegels *et al* 1995
HMW glutenin	Sequential extraction	Kipp *et al* 1996; Nicolas *et al* 1997
LMW glutenin	Sequential extraction, Isoelectric focusing	Larre *et al* 1997; Sissons *et al* 1998
Albumin, globulin	Sequential extraction	Prieto *et al* 1994
General procedures for the purification of wheat prolamins		Graveland *et al* 1980; Bietz, 1997

References

Bean, S.R. and Lookhart, G.L. 1998. Influence of salts and aggregation of gluten proteins on reduction and extraction of high molecular weight glutenin subunits of wheat. Cereal Chem. 75:75-79.

Bean, S.R., Bietz, J.A. and Lookhart, G.L. 1998. High-performance capillary electrophoresis of cereal proteins. J. Chromatog. A 814:25-41.

Bean, S.R., Lyne, R.K., Tilley, K.A., Chung, O.K. and Lookhart, G.L. 1999. A rapid method for quantitation of insoluble polymeric proteins in flour. Cereal Chem. 75:374-379.

Békés, F. and Gras, P. 1999. In vitro studies on gluten functionality. Cereal Foods World 44:580-585.

Bekkers, A., De Boef, E., Van-Dijk, A.-A. and Hamer, R.J. 1999. The central domain of high molecular weight glutenin subunits is water-soluble. J. Cereal Sci. 29:109-112.

Belton, P.S., Shewry, P.R., and Tatham, A.S. 1999. ^{13}C solid state nuclear magnetic resonance study of wheat gluten. J. Cereal Sci. 3:305-317.

Bietz, J.A. 1983. Separation of cereal proteins by reversed-phase high-performance liquid chromatography. J. Chromatog. 255, 219-238.

Bietz, J.A. 1984. Analysis of wheat gluten proteins by high-performance liquid chromatography. I. Bakers' Dig. 58:15-17.

Bietz, J.A. 1985. High performance liquid chromatography: how proteins look in cereals. Cereal Chem. 62:201-212.

Bietz, J.A. 1997. Recent advances in the isolation and characterization of cereal proteins. Cereal Foods World 24:199-202.

Bietz, J.A. and Wall, J.S. 1973. Isolation and characterization of gliadin-like subunits from glutenin. Cereal Chem. 50:537-547.

Chevallier, S., Colonna, P., Della Valle, G. and Lourdin, D. 1999. Structural modifications of biscuit doughs during baking-role of ingredients. In: "Biopolymer Science: Food and Non Food Applications." P. Colonna and S. Guilbert eds. INRA editions, Montpellier (France). pp 191-197.

Don, C., Lichtendonk, W.J., Plijter, J.J. and Hamer, R.J. 2002. Glutenin Macropolymer: a gel formed by glutenin particles . J. Cereal-Sci. in press.

Graveland, A. 1980. Extraction of wheat proteins with sodium dodecyl sulphate. Ann. Technol. Agr. 29:113-123.

Graveland,A., Bosveld, P., Lichtendonk, W.J., Moonen, H.H.E. and Scheepstra, A. 1980. Extraction and fractionation of wheat flour proteins. J. Sci. Food Agric. 33:1117-1128.

Gupta, R.B., Batey, I.L. and MacRitchie, F. 1992. Relationships between protein composition and functional properties of wheat flours. Cereal Chem. 69:125-131.

Gupta, R., Khan, K. and MacRitchie, F. 1993. Biochemical basis of flour properties in bread wheats. 1. Effects of variation in the quantity and size distribution of polymeric protein. J. Cereal Sci. 18:23-41.

Hargreaves, J., Meste, M. and Popineau, Y. 1994a. ESR studies of gluten-lipid systems. J. Cereal Sci. 19:107-113.

Hargreaves, J., Meste, M., Cornec, M. and Popineau, Y. 1994b. Electron spin resonance studies of wheat protein fractions. J. Agric. Food Chem. 42:2698-2702.

Kipp, B., Belitz, H.D., Seilmeier, W. and Wieser, H. 1996. Comparative studies of high M_r subunits of rye and wheat. I. Isolation and biochemical characterisation and effects on gluten extensibility. J. Cereal Sci. 23:227-234.

Larre, C., Popineau, Y. and Loisel, W. 1991. Fractionation of gliadins from common wheat by cation exchange FPLC. J. Cereal Sci. 14:231-241.

Larre, C., Nicolas, Y., Desserme, C., Courcoux, P. and Popineau, Y. 1997. Preparative separation of high and low molecular weight subunits of glutenin from wheat. J. Cereal Sci. 25:143-150.

Lupano, C.E. and Anon, M.C. 1987. Denaturation of wheat endosperm proteins during drying. Cereal Chem. 64:437-444.

Marion, D., LeRoux, C., Akoka, S., Tellier, C. and Gallant, D. 1987. Lipid-protein interactions in wheat gluten: a phosphorus nuclear magnetic resonance spectroscopy and freeze-fracture electron microscopy study. J. Cereal Sci. 5:101-115.

Nicolas, Y., Larre, C., and Popineau, Y. 1997. A method for the isolation of high M_r subunits of wheat glutenin. J. Cereal Sci. 25:151-154.

Nicolas, Y., Martinant, J.P., Denery, P.S. and Popineau, Y. 1998. Analysis of wheat storage proteins by exhaustive sequential extraction followed by RP-HPLC and nitrogen determination. J. Sci. Food Agric. 77:96-102.

Popineau, Y. 1985. Fractionation of acetic acid-soluble proteins from wheat gluten by hydrophobic interaction chromatography: evidence for different behaviour of gliadin and glutenin proteins. J. Cereal Sci. 3:29-38.

Popineau, Y. and Pineau, F. 1985 Fractionation of wheat gliadins by ion-exchange chromatography on SP Trisacryl M. Lebensm. Wiss. Technol. 18:133-135.

Popineau, Y. and Pineau, F. 1988. Changes of conformation and surface hydrophobicity of gliadins. Lebensm. Wiss. Technol. 21:113-117.

Prieto, J.A., Weegels, P.L. and Hamer, R.J. 1994. Functional properties of low M_r wheat proteins: I. Isolation, characterization and comparison with other reported low M_r wheat proteins. J. Cereal Sci. 17:203-220.

Singh, N.K., Donovan, G.R., Batey, I.L. and MacRitchie, F. 2000. Use of sonication and size-exclusion high-performance liquid chromatography in the study of wheat flour proteins. I. Dissolution of total proteins in the absence of reducing agents. Cereal Chem. 67:150-161.

Sissons, M.J., Békés, F. and Skerritt, J.H. 1998. Isolation and functionality testing of low molecular weight glutenin subunits. Cereal Chem. 75:30-36.

Tamas, L., Békés, F., Greenfield, J., Tatham, A.S., Gras, P.W., Shewry, P.R. and Appels, R. 1998. Heterologous expression and dough mixing studies of wild-type and mutant C hordeins. J. Cereal Sci. 27:15-22.

Weegels, P.L., Marseille, J.P., Bosveld, P. and Hamer, R.J. 1995. Large-scale separation of gliadins and their bread-making quality. J. Cereal Sci. 20:253-264.

Weegels, P.L., Hamer, R.J., and Schofield, J.D. 1996. Functional properties of wheat glutenin. J. Cereal Sci. 23:1-18.

Weegels, P.L., Hamer, R.J. and Schofield, J.D. 1997. Depolymerisation and re-polymerisation of wheat glutenin during dough processing. II. Changes in composition. J. Cereal Sci. 25:155-163.

Wrigley, C.W., Andrews, J.L., Békés, F., Gras, P.W., Gupta, R., MacRitchie, F. and Skerritt, J.H. 1999. Protein-protein interactions-essential to dough rheology. In Interactions: The keys to Cereal Quality. R.J.Hamer and Hoseney,R.C., eds. Am. Assoc. Cereal Chemists, Inc., St. Paul, MN, USA. 17-46.

Chapter 3

Electrophoresis of Wheat Gluten Proteins

Khalil Khan and Gloria Nygard
Department of Cereal and Food Sciences, North Dakota State University
Fargo, ND, USA

Norberto E. Pogna and Rita Redaelli
Instituto Sperimentale per la Cerealicoltura, Rome, Italy

Perry K.W. Ng
Department of Food Science & Human Nutrition, Michigan State University
East Lansing, MI, USA

Roger J. Fido and Peter R. Shewry
Long Ashton Research Station, Department of Agricultural Sciences
University of Bristol, Long Ashton, Bristol BS 41 9AF, UK

Electrophoresis is one of the most important methods used for analyzing proteins. The versatility of the technique allows it to be applied to proteins isolated from a diverse array of plant or animal sources and to samples representing many types and sizes of proteins. The samples may be isolated by a variety of extraction methods, the proteins may be in their native form or reduced, and the electrophoresis may be performed under acidic, basic, or neutral conditions, and may utilize several mechanisms of separation. This chapter will describe several of the main types of electrophoresis used by cereal chemists—sodium dodecylsulfate-polyacrylamide gel electrophoresis (SDS-PAGE), multi-stacking-sodium dodecylsulfate-polyacrylamide gel electrophoresis (MS-SDS-PAGE), lactic acid-polyacrylamide gel electrophoresis (A-PAGE), isoelectric focusing (IEF), two-dimensional (2-D) methods, and western blotting. Related techniques such as capillary electrophoresis (CE) are described in other chapters.

Sodium Dodecylsulfate-Polyacrylamide Gel Electrophoresis

SDS-PAGE based on the method of Laemmli (1970) is the most widely used system of electrophoresis applied to the analysis of proteins. Many reference books (e.g. Hames and Rickwood, 1990; Bollag *et al* 1996; Rosenberg, 1996) and review articles (e.g. Blackshear, 1984; Garfin, 1990) provide detailed step-by-step protocols for the Laemmli method, including reagent preparation, gel polymerization, sample preparation, gel running conditions, and protein staining, so detailed laboratory procedures are not included here. The manufacturers of electrophoresis equipment and supplies (e.g. Amersham Biosciences www.amershambiosciences.com and Bio-Rad www.bio-rad.com) are excellent sources of general electrophoresis information as well as technical information concerning their products. Ready-to-use gels and reagents are available from many suppliers.

General Principles of SDS-PAGE

Polyacrylamide gels provide a versatile, porous matrix that can be used to separate proteins on the basis of charge and/or size. Pore sizes in acrylamide gels vary inversely with the percent acrylamide and depend on the ratio of acrylamide to bis-acrylamide (the cross-linker) as well as on the conditions of polymerization. The acrylamide concentration and the choice of buffers and other additives (e.g. detergents and reducing agents) give the scientist flexibility in the separation of complex molecules.

Sodium dodecylsulfate (SDS) is an anionic detergent that readily forms a complex with proteins. SDS disrupts hydrogen bonds, blocks hydrophobic interactions, and substantially unfolds the protein molecules while conferring them with a net negative charge. Under electrophoretic conditions, ionic species travel through the gel matrix, with larger molecules being retarded more than small molecules. The result of this sieving action is that migrating proteins are separated by molecular size and their apparent molecular weights can be determined by comparison with protein standards of known molecular weights.

The Laemmli method (1970) uses a discontinuous buffer and gel system—i.e. uses different pH and acrylamide concentrations for the separating and stacking gels. This leads to concentration (stacking) of proteins into a sharp band at the interface of the gel layers, and the proteins then continue as sharp bands as they travel through the separating gel.

Chemicals and Reagents for SDS-PAGE

SAFETY PRECAUTIONS: Acrylamide and N,N'-Methylene-bis-acrylamide are neurotoxins. Users should use good laboratory practices, work in a well-ventilated area, wear proper personal protective equipment, and refer to Material Safety Data Sheets for further information. General guidelines for use of acrylamide include the following precautions. Gloves should be worn to prevent contact of acrylamide or bis-acrylamide powders and solutions with skin. Inhalation of acrylamide dust must be avoided and spills cleaned up immediately. Polymerization renders acrylamide nontoxic, but residual reagents should be rinsed away before handling gels with bare hands. Any unused acrylamide monomer solutions should be polymerized in a beaker and then the gel discarded.

All electrophoresis reagents should be prepared using the highest purity chemicals available in order to avoid contaminants that might cause inconsistent gel polymerization or changes in separation characteristics.

Apparatus

Electrophoresis equipment is available from several manufacturers, and a typical system includes a gel forming apparatus (i.e. glass plates, spacers, rubber gaskets, slot formers, and casting stand), electrodes, buffer chambers, heat exchanger, temperature-controlled circulating water bath, and power supply. The apparatus may be configured to hold gels in a vertical or horizontal position, and may be designed to accommodate standard size gels, mini-gels, or tubular gels. The choice of what size or type of gel to use for a particular application is dependent on the complexity of the protein mixtures to be separated and on what subsequent steps of analysis (e.g. 2-D gels) to be performed. An electrophoresis unit that is well suited for use with SDS-PAGE, MS-SDS-PAGE gels, and A-PAGE is the Hoefer Model SE 600 Standard Vertical Slab Unit (Amersham BioSciences) with 18x16 cm glass plates.

Procedures

Preparation of Gels for SDS-PAGE

Reagent Solutions for Laemmli SDS-PAGE: Acrylamide/Bis-acrylamide Solution (30%T, 0.8%C); 1.5M Tris-HCl, pH 8.8 Buffer for use in separating gels; 0.5M Tris-HCl, pH 6.8 Buffer for use in stacking gels; 10% (w/v) SDS; and freshly prepared 10% (w/v) Ammonium Persulfate (APS).

A typical separating gel (e.g. 12%T) would be prepared by combining measured volumes of the above reagent solutions along with water. N,N,N',N'-Tetramethylethylenediamine (TEMED) is added to initiate polymerization, the solution transferred into a pre-assembled glass plate sandwich (0.75-1.5mm thick), and then overlaid with water to exclude air during the polymerization process. After the separating gel has polymerized, the top surface is rinsed, and a stacking gel solution of lower acrylamide (e.g. 5%T) concentration is prepared and allowed to polymerize around the sample comb.

Preparation of Samples and Standards for SDS-PAGE

Reagents for Sample Preparation: Non-reducing Sample Buffer (0.0625M Tris-HCl, pH 6.8, 20% glycerol, 2% SDS, 0.0002% Pyronin-Y or Bromophenol Blue); and Reducing Sample Buffer (5% 2-Mercaptoethanol or 1% Dithiothreitol (DTT) in Non-reducing Sample Buffer).

Weigh samples into 1.5mL microcentrifuge tubes, add Sample Buffer, cap the tubes, and extract for 1-2 hours. Suggested sample sizes and extraction volumes are: 50 mg flour or meal/1 mL; 10 mg protein/1 mL; or a single kernel (crushed)/0.5 mL. To facilitate extraction, a multi-tube mixer (e.g. Eppendorf Thermomixer, Brinkman Instruments, Inc., Westbury, NY) may be used at moderate temperature (e.g. 30-60C). The tubes should be subjected periodically to additional vortexing at high speed to thoroughly resuspend the samples in extraction solution. After the extraction is completed, the samples are placed in a nearly boiling water bath (95C) for 5 minutes, allowed to cool, and then centrifuged.

Reducing or Non-reducing Sample Buffer may be used to extract proteins directly from ground grain samples, or one of the buffers may be combined (e.g. 1:1, v/v) with other types of sample extracts. In cases where sample extracts contain a large percentage of organic solvent (e.g. 50% alcohol), addition of solid sucrose (to approx. 10% w/v) may be necessary to increase the sample density before loading an aliquot of the sample onto the gels for electrophoresis. A molecular weight standard (e.g. Bio-Rad SDS-PAGE MW Standard) should be loaded in one lane of each electrophoresis gel.

Running SDS-PAGE Gels

WARNING: High voltage is a hazard during the electrophoretic run. The Electrode Running Buffer (Tank Buffer) for Laemmli SDS-PAGE is 0.025M Tris base + 0.192M glycine + 0.1% SDS at pH 8.3. Assemble the electrophoresis apparatus (buffer reservoirs, water-cooled heat exchanger, gel sandwiches, and electrodes) according to the manufacturer's instructions. The direction of migration of the negatively charged protein/SDS complex is toward the anode. In order to minimize heat-related band distortion and maintain a constant rate of migration throughout the run, vertical slab gels are usually run using low current for several hours or overnight.

The electrophoresis run is complete when the marker dye (Pyronin-Y or Bromophenol Blue) has traveled to the bottom of the gels. Turn off the power supply and cooling water, disassemble the apparatus, and place each gel sandwich into a separate labeled tray. Disassemble the gel sandwiches, gently loosen the gels from the glass plates, and trim off a portion of the fringe-like row of sample well dividers at the top of the gels. The gels are now ready to be fixed and stained.

Staining of Proteins in SDS-PAGE Gels

Many staining methods are available for visualizing proteins on electrophoresis gels, and the methods vary in sensitivity, selectivity, convenience, and color stability. Two of the most widely used gel stains are Coomassie Brilliant Blue-R-250 (CBB-R) and Coomassie Brilliant Blue-G-250 (CBB-G). Other protein staining methods include organic dyes, silver and copper stains, fluorescent stains, and some selective and/or reversible stains. These methods have been reviewed recently (Patton, 2001) as well as over the years in Methods in Enzymology (Wilson, 1983; Cutting, 1984; Gander, 1984; Reisner, 1984; Merril *et al* 1984; Merril, 1990).

Staining Method I—CBB-R in Acetic Acid/Methanol/Water

Coomassie Brilliant Blue-R (CBB-R) dissolved in acetic acid/methanol/water solution is the most widely used staining method for proteins in SDS-PAGE gels (Rosenberg, 1996). The proteins are fixed as well as stained when the gels are immersed in Staining Solution (0.25% CBB-R in 10% Acetic acid + 50% Methanol + 40% Water) and gently agitated for 1 hour. When the Staining Solution is poured off, the gels appear to be dark blue throughout. Excess stain is removed from the gel background by immersing the stained gel in Destaining Solution (7% Acetic acid + 20% Methanol + 73% Water). When the Destaining Solution becomes highly colored, it is replaced with fresh solution. The destaining is discontinued when the background of the gel becomes nearly clear while the protein bands maintain color.

Staining Method II—CBB-G Colloidal Particle Method

Coomassie Brilliant Blue-G (CBB-G) is typically used as a colloidal suspension of dye particles that do not penetrate deeply into the gel, but rather form a colored complex with proteins near the surface of the gel (Neuhoff *et al* 1988 and 1990). The CBB-G colloidal method has the following advantages: higher sensitivity than the dissolved CBB-R method; the capability of staining certain proteins, which are not detected with CBB-R (Nygard and Khan, *unpublished*); and the maintenance of stable protein band color during storage of gels.

Reagent solutions for the Neuhoff *et al* (1988) method include: Fixation Solution—12% Trichloroacetic Acid (TCA); Stock Stain Solution—0.1% CBB-G + 2% Phosphoric Acid + 10% Ammonium Sulfate; Working Stain Solution—80% Stock Stain Solution + 20% Methanol; and Gel Storage Solution—20% Ammonium Sulfate. The stain solutions are not filtered and should be shaken thoroughly just before use to resuspend the colloidal dye particles. The separated proteins are fixed by immersing and gently agitating the gels in 12% TCA for 1 hour. The fixing solution is poured off and replaced with Working Stain Solution, and the gels are stained for several hours or overnight. The protein bands gradually appear stained against a relatively clear background. The gels are then rinsed briefly (< 1 minute) with methanol-water (25:75, v/v) and stored in 20% $(NH_4)_2SO_4$ solution.

Documentation of Gel Images and Storage of Stained Gels

Images of gels may be captured using a photography system (film or digital) or an imaging densitometer. A scanner having both reflectance and transmittance capability allows for flexibility in accurately documenting and quantitating the protein bands on the gel images. Most gels may be stored for several months by simply placing them inside plastic zip-lock type bags and keeping them wetted with a small amount of destaining solution or storage solution. Gels may also be dried before storage, and gel-drying equipment is available from several manufacturers.

Examples of Applications of SDS-PAGE

SDS-PAGE is used routinely in the study of wheat proteins. The technique is used to compare proteins isolated by various extraction methods, to screen inherited protein patterns in plant breeding, to determine HMWGS or LMWGS composition, and to look for changes in proteins which occur during dough mixing and baking or during storage of food products.

Tips and Method Modifications for SDS-PAGE

The most common variables found in the SDS-PAGE methods published by different cereal science research groups are the choices of acrylamide and bis-acrylamide concentrations. As %T and %C are varied, the pore sizes of the gels change such that different MW ranges can be separated and some proteins become better resolved. An example of using this gel property to advantage is in the determination of HMWGS composition where certain pairs of subunits (e.g. 2 and 2*) may overlap on a 10% gel, but be resolved using a 5% gel (Lukow et al 1989; Rogers et al 1989). The effect of changing gel pore sizes, though, is not the same for all proteins— i.e. on these same gels, one might observe that another pair of subunits (e.g. 2* and 5), which had been separated at 10%, might now be poorly resolved at 5%. Finding the best acrylamide concentrations for a particular separation may require testing several gel concentrations since even small differences in the reagent concentrations can affect gel polymerization and the resolution of complex samples. Other common method variations include the use of higher or lower pH buffers (Khan and Bushuk, 1977; Zhen and Mares, 1992) or the use of higher (e.g. 17%T) acrylamide concentrations for the gels (Ng and Bushuk, 1987).

Modified SDS-PAGE methods that incorporate urea into the gel system have shown good band resolution, but result in changed relative mobilities for certain HMWGS and other proteins compared to the Laemmli system (Mackie et al 1996; Lafiandra et al 1999; Gianibelli et al 2001). Gradient gels may also be useful for resolving complex samples having a wide MW range, but casting reproducible gradient gels is difficult.

Gel Preparation

Gels prepared at different percentages or thicknesses will vary in physical texture, sample resolution, and the extent of background staining. A thickness of 1.5mm is commonly used, but thinner gels (e.g. 0.75mm) provide advantages including better resolution of certain proteins (Huang and Khan, 1998) and improved sensitivity for analyzing low concentrations of proteins, especially if colloidal CBB-G is used for staining. The advantages of thin gels, though, must be weighed against certain disadvantages such as reduced physical strength, which can cause difficulty in handling. Preparative gels are often made 3mm thick.

Sample Preparation and Gel Loading

Distortion of protein bands on gels may be caused by sample overloading, by loading unequal sample volumes into adjacent lanes, and by lateral migration during electrophoresis. These types of distortion are minimized when sample loading volumes are kept uniform across the gel (i.e. by adding extra buffer to equalize sample volumes and loading blank sample buffer into empty sample wells), and when the loading of reduced and non-reduced samples side-by-side is avoided.

Alkylation of reduced sample proteins prior to electrophoresis has been reported to improve the resolution of glutenin subunits (Gupta and MacRitchie, 1991).

Gel Staining

The concentrations of ingredients in staining and destaining solutions vary between published methods, and these differences may affect protein fixation as well as the rate of staining or color stability. When using dissolved CBB-R stain, increased methanol content will fix proteins rapidly and facilitate rapid penetration of stain through the gel, but high methanol concentrations in destaining solutions may cause loss of color from protein bands (Neuhoff *et al* 1990).

According to Neuhoff *et al* (1988) the colloidal properties of CBB-G are enhanced under acidic conditions, at higher ammonium sulfate concentrations, and at lower methanol concentrations. In our laboratory we have obtained good CBB-G staining results with minimal background color by modifying the Neuhoff procedure to use 15% ammonium sulfate and 15% methanol when gels are being stained overnight (Nygard and Khan, *unpublished*). Quantitation of protein bands in gels has been shown to be reproducible for thin (e.g. 0.75mm) CBB-G stained gels (Huang and Khan, 1998; Zhu and Khan, 1999).

Used staining solutions may be saved and reused on future gels, but the evaporation of volatile ingredients, the loss of colloidal stain particles, and/or the accumulation of extraneous reagents (e.g. SDS) in the used staining solution will eventually reduce its effectiveness and require the preparation of fresh staining solution. If stained protein bands fade during storage, the gels can usually be restained with either the same method or using another staining method. If gels dry out or shrink in storage, soaking for a few hours in an appropriate destaining or storage solution can usually rehydrate them.

Multi-Stacking-Sodium Dodecylsulfate-Polyacrylamide Gel Electrophoresis (MS-SDS-PAGE)

The same equipment and reagents are used for MS-SDS-PAGE as for SDS-PAGE. Procedures are modified to include the use of only non-reducing solutions for sample preparation and the incorporation of several stacking layers in order to fractionate very large proteins. Polymeric proteins are trapped at the interfaces of the stacking gel layers since the successive layers have sharp changes in pore sizes.

Procedures

Preparation of Multi-Stacking Gels
In the MS-SDS-PAGE method of Khan and Huckle (1992), the separating gel (14% acrylamide) is poured to a height 5cm from the top of the glass plates and overlaid as usual with water in order to form a sharp top boundary when it polymerizes. A series of five stacking gels (equal to 12, 10, 8, 6, and 4% acrylamide) are then successively polymerized above the 14% separating gel. Each stacking layer is made 0.5cm high, and the sample wells are formed in the 4% layer. To facilitate formation of uniform stacking layers, a slot former (which fits very tightly between the glass plates) is held upside down (i.e. with the long flat side pointing downward), and pushed into the stacking gel solution at an angle while being careful to exclude air bubbles. The slot former is pushed downward until it is centered from side to side, oriented horizontally, and positioned 0.5cm above the previously polymerized gel. Use of an inverted slot former in this way provides a smooth horizontal surface for the gel to polymerize against and avoids the risk of diluting the stacking gel solutions (a problem which could be encountered if the technique of overlaying each layer with water was used to exclude air during polymerization). Approximately 45 minutes is allowed for polymerization of each stacking layer, then the slot former is pulled out, excess gel that has polymerized along the walls above the gel layer is scraped out, and the gel debris is rinsed away. The gel sandwich is drained, and absorbent paper is used to blot any remaining water droplets from between the glass plates. A small portion (about 1 mL) of the next stacking gel solution may be used to rinse the surface of the previously polymerized gel (and then discarded) before pouring in the remaining solution and pushing in the slot former for the next layer to be formed.

Characterization of Proteins Entrapped at MS Gel Layer Interfaces and Examples of MS-SDS-PAGE Applications
After electrophoresis, MS gels may be stained to visualize the protein separation, or the gels may be cut into pieces for subsequent analyses. MS gels stained with dissolved CBB-R show protein trapped at each gel layer interface, a smear of non-reduced proteins throughout the gel lane, and

some individual protein bands in the separating gel. When used as a preparative technique, thicker (e.g. 3mm) gels are run, then the stacking gel interface sections are excised, and the polymeric proteins are reduced and extracted for characterization by SDS-PAGE (Huang and Khan, 1997a) or by another method.

Procedure: Place one or two excised gel pieces into a 1.5mL tube, crush, and extract with 0.5mL Reducing Sample Buffer at 60C for 2 hours. Heat at 95C for 5 minutes, cool, and centrifuge. Load 50μL onto a 0.75mm SDS-PAGE gel, electrophorese, and stain with colloidal particle CBB-G (Neuhoff *et al* 1988).

Experiments may include looking for differences in the HMWGS composition of the proteins that were trapped at each MS interface, comparing the size distribution of polymeric proteins from different cultivars or classes of wheat, and looking for differences in the polymeric proteins solubilized by different extraction methods. The polymeric glutenin proteins isolated from MS gels have been reported to show variations in the proportions of HMWGS as well as of other proteins. These differences have been observed in comparing protein content at different growth stages (Zhu and Khan, 1999) or from plants subjected to various environmental conditions (Zhu *et al* 1999; Zhu and Khan, 2001). Changes in the polymeric protein fractions have also been correlated with certain stages of dough mixing and development, as well as baking and storage (Huang and Khan, 1996, 1997b, 1997c; Borneo and Khan, 1999).

Lactic Acid-Polyacrylamide Gel Electrophoresis (A-PAGE)

General Principles of A-PAGE

A-PAGE (based on the methods of Jones *et al* 1959; Woychik *et al* 1961; Bushuk and Zillman, 1978; Lookhart *et al* 1982; and Khan *et al* 1983 and 1985) is used primarily for the separation of prolamin proteins for the purpose of cultivar identification (Khan, 1982). The method is also used for the separation of other monomeric proteins including albumins and globulins.

Chemicals and Reagents for A-PAGE

Reagent Solutions for A-PAGE: Electrode Running Buffer (Tank Buffer)—0.25% Aluminum Lactate, pH adjusted to 3.1 with Lactic Acid; Acrylamide/Bis-acrylamide Stock Solution (7.25 %T, 3.45 %C)—7.0% Acrylamide + 0.25% Bis-acrylamide + 0.024% Ascorbic Acid + 0.0004% Ferrous Sulfate in Tank Buffer; Polymerization Catalyst Solution—3% Hydrogen Peroxide (H_2O_2); Sample Extraction Solution—70% Ethanol + 30% Water; Sample Buffer—10g sucrose + 0.2g Methyl Green + 18mL Tank Buffer.

Procedures

Preparation of A-PAGE Gels and Samples

Gels: Combine appropriate volumes of chilled acrylamide and H_2O_2 solutions (2mL: 1μL; v/v). Immediately pour into glass plate sandwich, and insert slot former. A-PAGE gels polymerize rapidly and are usually cast 1.5mm thick as a continuous gel without a stacking layer.

Samples: Weigh 100mg flour or meal, and extract with 300μL70% ethanol for 1hour. Centrifuge, transfer an aliquot of supernatant to a clean tube, combine 1:1 with Sample Buffer, and load 5μL. For analysis of a single kernel (crushed), extract with 200μl and load 8-10μL. Extracts of known cultivars should be run on each gel and used for direct comparison to unknown samples.

Running and Staining A-PAGE Gels

Connect the electrodes in the "reversed" direction compared to SDS-PAGE (i.e. for migration from anode to cathode), and run the gels until the slower (purple) of the two Methyl Green tracking dyes reaches the bottom of the gel. The use of aluminum lactate tank buffer results in high running voltage (e.g. ~350v for 50mA constant current).

The colloidal particle CBB-G staining method (see above) works well for A-PAGE gels, having good sensitivity and color stability for the stained bands. Another common staining procedure for A-PAGE gels uses a single fixing and staining solution (prepared fresh daily by combining 15mL of Stock Stain Solution (1.0% CBB-R in 95% Ethanol) with 300mL of 12% TCA solution. The gels are stained for several hours or overnight, followed by destaining in 12% TCA (Bushuk and Zillman, 1978; Lookhart *et al* 1982). This second method has good sensitivity, but poor color stability during gel storage in TCA solution (Wilson, 1983).

Note: A-PAGE gels are fragile (i.e. less elastic than SDS-PAGE gels) and can tear easily. It is recommended that A-PAGE gels be supported by a glass or plastic sheet when handled.

Examples of Applications, Tips and Modifications for A-PAGE

A-PAGE is used primarily for gliadin separations in variety identification, but it is also useful for other proteins. Albumins and globulins travel rapidly through A-PAGE gels, eluting between the two Methyl Green marker dyes, such that the electrophoresis must be stopped when the first dye (green) reaches the bottom of the gel. A-PAGE also can provide useful data (complimentary to the SDS-PAGE gel patterns) for comparison of proteins isolated by various non-reducing or multi-step extraction methods (Kim and Bushuk, 1995; Dupuis *et al* 1996).

An alternative to aluminum lactate A-PAGE for variety identification is a method utilizing an acetic acid-glycine tank buffer with urea incorporated in the gel recipe (Cooke, 1992).

Two-Dimensional IEF × SDS-PAGE and NEPHGE × SDS-PAGE

General Principles of the Techniques

Mature wheat grain contains several types of proteins including storage proteins (gliadins and glutenins), albumins, and globulins (enzymes, membrane proteins, etc.). The most satisfactory way of separating these protein types is by using a combination of two high-resolution procedures developed by O'Farrell in the mid-seventies (O'Farrell, 1975). These are two-dimensional electrophoretic techniques in which proteins are fractionated in one dimension on the basis of charge and in the second dimension by size. Because denaturants are used in both dimensions, the procedures resolve the proteins in terms of their polypeptide subunits. Both techniques involve SDS-PAGE in the second dimension, but differ from each other in the first dimension. The first technique (O'Farrell, 1975) uses isoelectrofocusing (IEF) in the first dimension, whereas the second one (O'Farrell *et al* 1977) utilizes nonequilibrium pH gradient electrophoresis (NEPHGE) to additionally resolve those proteins with isoelectric points greater than 7.5.

IEF is a method to separate amphoteric biological molecules, such as proteins, based on their isoelectric points (pI) by using a pH gradient established between two electrodes and stabilized by carrier ampholytes. IEF is therefore an equilibrium technique in which the effects of diffusion are reduced. However, during the IEF run, the ampholytes first focus in the basic region at the top of the gel, and then base hydrolysis of the gel causes fixed charges to form in that region, which causes the basic ampholytes and proteins to move upwards and out of the gel into the buffer tank. This phenomenon, known as "cathodic drift," causes loss of the basic ampholytes and proteins, but can be avoided with the use of NEPHGE to fractionate the basic proteins. This latter technique requires that the anode and cathode solutions be reversed so that the sample is applied to the anodic end. Separation is then carried out for a shorter time than IEF in order to obtain a transient state of focusing pattern.

The O'Farrell procedures can routinely resolve over 1000 polypeptides on a single gel because they separate proteins by two different parameters with very high resolutions, yielding a combined resolution better than 0.1 charge units and about 1000 daltons (Anderson and Anderson, 1977).

Proteomics, the modern protein separation science, largely relies on two dimensional IEF × SDS-PAGE or NEPHGE × SDS-PAGE separation of proteins, the results of which are read and mapped using powerful computer algorithms (Wilkins *et al* 1997).

High-resolution two-dimensional gel electrophoresis can be used to detect and purify hundreds of polypeptides from a single sample simultaneously. Using computerized systems for analyzing 2-D images and constructing spot databases, it is possible to plan and assemble integrated bodies of information describing the appearance and regulation of thousands of protein gene products. Identification of every spot can be achieved by western blotting with subsequent immunodetection, mass spectrometry, or Edman amino acid microsequencing of isolated polypeptides. Furthermore, the World Wide Web has become the standard interface for distributing the 2-D protein databases, and tools have been developed to compare 2-D images across the Internet.

Two Dimensional IEF × SDS-PAGE

Since its introduction in 1975, two-dimensional IEF × SDS-PAGE has become the major technique used for research on expression and post-translational modifications of many types of proteins, including endosperm proteins of wheat. Wheat gliadins, HMW glutenin subunits and the minor neutral and acidic LMW subunits of glutenin were first fractionated by this technique more than 20 years ago using carrier ampholyte-generated pH gradients in gel rods (Brown *et al* 1979; Brown and Flavell, 1981; Holt *et al* 1981).

First dimension fractionation by IEF in tubes

The first dimension fractionation by isoelectric focusing (IEF) is performed in individual gel rods.

Materials

1. Acrylamide and N,N'-Methylenebisacrylamide. Electrophoresis grade acrylamide without contaminant metal ions is sold by several companies. A 30%-acrylamide stock solution containing 5.68 g acrylamide and 0.32 g N,N'-methylenebisacrylamide (bisacrylamide) should be prepared in 20 mL deionized water. Gloves and a mask must be used when weighing neurotoxic acrylamide and bisacrylamide. Check that the pH of the solution is ≤ 7.0 and store the solution for up to a few months in dark bottles at 4°C.
2. Ammonium Persulfate (APS). A small volume (10 mL) of 10% (w/v) ammonium persulfate solution in deionized water should be prepared on the day of use. It provides the free radicals for polymerization of acrylamide and bisacrylamide.
3. TEMED (N,N',N'-Tetramethylethylenediamine). Electrophoresis grade TEMED is sold by many companies. It catalyzes the formation of free radicals from APS.
4. Nonidet P-40. A small volume (30 mL) of 10% (w/v) nonidet P-40 detergent in deionized water should be prepared and may be stored at room temperature for up to 1 month.
5. Carrier Ampholytes. Several manufacturers sell special grades of carrier ampholytes with different pH ranges. A good fractionation of acidic proteins can be obtained using 0.80 mL of a mixture containing 0.163 mL each of Ampholyte Solution (AS) pH 8.0-9.5, AS pH 7.0-9.0, AS pH 5.0-7.0, AS pH 4.0-6.0, and 0.150 mL of AS pH 3.5-10.0.

Alternatively, a good overall separation can be achieved using 0.80 mL of a mixture containing AS pH 8.0-9.5 (0.1625 mL), AS pH 7.0-9.0 (0.1625 mL), AS pH 5.0-7.0 (0.325 mL), and AS pH 3.5-10.0 (0.150 mL). It is best to use a different syringe for each ampholyte solution to avoid cross contamination.

6. Urea. Dissolve 24.0 g urea in 32.25 mL boiling deionized water, then divide into 1.0 mL aliquots and store at –20°C. This 8 M urea solution will be used for overlaying the gel rods.

7. Anode Buffer (lower). This buffer contains 15 mM orthophosphoric acid (H_3PO_4) and is made by dissolving 0.3 mL of 85% orthophosphoric acid in 299.7 mL of deionized water. This solution should be prepared on the day of use.

8. Cathode Buffer (upper). This buffer contains 20 mM NaOH and can be made by dissolving 0.8 g of NaOH in deionized water to a final volume of 1L. This buffer should be prepared fresh from solid NaOH. (However, it can also be stored under vacuum.)

9. SDS Equilibration Solution. This buffer is used for equilibrating gel rods prior to electrophoresis in the second dimension (SDS-PAGE). It can be made by dissolving 49.0 g glycerol, 4.3 g Tris, and 11.5 g SDS in about 300 mL of deionized water. This buffer should be adjusted to pH 6.8 with 6M HCl before making the volume up to 475 mL.

Electrophoresis Unit

The use of gel rods requires IEF to be carried out in vertical electrophoretic apparatuses. These are sold by many companies including Amersham Biosciences, Bio-Rad, and Fluka. Various sizes of glass tubes can be used with the different electrophoresis units. The protocol described here is for glass tubes 90 mm in length with 5 mm internal diameter, each holding about 2 mL of gel solution.

Method

1. It is convenient to make up the gel rods in lots of eight. Before use, glass tubes should be soaked for 10 min in Kodak Photoflo 600 diluted 1:600, and dried in an oven.

2. Seal the bottoms of the glass tubes with several layers of Parafilm, mark them 1.5 cm from the top and set them up in a holder.

3. For 8 rods, 16 ml of gel solution should be prepared by adding 8.25 g urea, 2mL of 30% acrylamide stock solution (see Materials, #1), and 0.8 mL of carrier ampholytes (see Materials, #5) to 3.75 mL of deionized water in a vacuum flask. Degas and add 3 mL of 10% Nonidet P-40 solution (see Materials, #4); dissolve urea by immersing the vacuum flask in warm water, assisting dissolution by moderate swirling. Now add 18 μL of 10% APS solution (see Materials, #2) and 12 μL TEMED. Without delay, swirl the mixture and proceed to the next step.

4. Pour the acrylamide solution into the glass tubes to 1.5 cm from the top using a pipette or a syringe fitted with thin tubing. Rods should be flicked hard when half full to dislodge air bubbles trapped at the bottom. Using a syringe, carefully overlay the acrylamide solution with a small amount of deionized water to prevent oxygen from diffusing into the gel and inhibiting polymerization. Cover the tops of the tubes with Parafilm and leave to polymerize for at least 2 h.

5. After polymerization is complete, pour off the overlay, and wash the top of each gel several times with deionized water to remove any unpolymerized acrylamide. Then remove the Parafilm from the bottom of each glass tube and replace with dialysis tubing, being careful to avoid trapping air bubbles in the bottoms of the gel tubes. Roll a rubber band onto the gel tube to hold the dialysis tubing in place.

6. Mount the gel tubes in the electrophoresis apparatus with the Anode Buffer in the bottom (lower) reservoir.

7. Overlay each gel with 8M urea solution (see Materials, #6), filling half the gap at the top of the rod. Fill the remaining gap with Cathode Buffer and add Cathode Buffer to the top reservoir. The tops of the gel rods must be fully submerged in the cathode solution.

8. Attach the electrophoresis unit to an electric power supply with the positive electrode connected to the bottom reservoir. Prefocus at 200 V for 30 min, 300 V for 30 min, and finally 400 V for 30 min. Turn off the power supply, pour off the cathode buffer, and replace.

9. Load the samples through the cathode buffer using a microsyringe and run at 400 V for 15 h and then at 800 V for 1 h.

10. Turn off the power supply, pour off the Cathode Buffer, and remove the glass rods from the apparatus. Note the orientation of each gel rod. Force the gel rods out of the glass tubes by air pressure from a syringe connected with rubber tubing and exude them into a sample tube (e.g., test tube) containing 23.75 mL of SDS Equilibration Solution. Add 1.25 mL of β-mercaptoethanol and place in a water bath at 35°C for 2 h before loading onto the second dimension gel. Gels can also be stored frozen before or after equilibration.

Second dimension fractionation by SDS-PAGE

Fractionation in the second dimension is SDS-PAGE with a discontinuous buffer system (see section above on SDS-PAGE). For this dimension, a slab electrophoresis apparatus is used. Many companies, including Bio-Rad, Amersham Biosciences, and Fluka, produce electrophoresis units for 2-D fractionations. These apparatuses have a U-shaped trough molded into the upper buffer chamber or beveled glass plates to hold the first dimension gel tube.

1. Glass plates with 1.5 mm spacers should be used. Plates are set vertically as usual and marked 1.5 cm from the top. Fill the plates to the 1.5 cm mark with the main (separating) gel solution (10%

acrylamide is appropriate for endosperm proteins), overlay with deionized water, and allow to set for at least 1 h.

2. Pour off the water overlay and wash the top of the gel several times with deionized water to remove any unpolymerized acrylamide, and then overlay with stacking gel solution.

3. Pour the stacking gel solution into the glass plates, leaving a 3 mm space from the top and overlay with deionized water. After polymerization is complete (30 min), pour water overlay off and rinse twice with deionized water. Then fill the gap between the top of the stacking gel and the top of the plate with hot 1% solution of agarose. Allow to solidify. The agarose solution can be made by dissolving 0.29 g agarose in 25 mL deionized water and 3.8 mL of 1M Tris, pH 6.8. Add enough bromophenol blue to color the solution and boil under reflux for 5 min.

4. Mount the gel in the electrophoresis apparatus and load the first dimension gel rod onto the second dimension gel and fix the gel rod in place with hot agarose excluding as many air bubbles from under the gel rod as possible.

5. Set the apparatus up for electrophoresis and run the gel until 1 h after the bromophenol blue reaches the bottom of the resolving gel. Then turn off the power supply, and fix and stain the second dimension gel as described previously (see section above on SDS-PAGE) or use it to establish a western blot (see section below on western blotting).

Two-Dimensional NEPHGE × SDS-PAGE

Since its introduction in 1977 (O'Farrell *et al* 1977), two-dimensional NEPHGE × SDS-PAGE has become the major technique used to resolve basic proteins with pI's above 7.5, including ribosomal and nuclear proteins with pI $\geq$ 10.0. Holt *et al* (1981) first applied this technique to fractionation of storage proteins from grain of wheat, whereas the two-dimensional NEPHGE × SDS-PAGE fractionation of basic LMW glutenin subunits was published in 1983 (Jackson *et al* 1983).

First dimension fractionation by NEPHGE in tubes

The first dimension fractionation by non-equilibrium pH gradient electrophoresis (NEPHGE) is performed in individual gel rods.

Materials

Acrylamide, N,N'-Methylenebisacrylamide, Ammonium Persulfate, TEMED, 10% Nonidet P-40 Solution, Anode Buffer, Cathode Buffer, and SDS Equilibration Solutions are the same as for IEF gels. In addition, the following solutions should be prepared:

1. 9.5 M Urea. Dissolve 5.71 g urea, 0.05g SDS, 0.2 mL β-mercaptoethanol, and 1 mL glycerol in deionized water to a final

volume of 10 mL. This solution should be prepared fresh immediately before use.

2. Lysis Buffer. This buffer contains 1.125 g of urea, 0.1 mL of β-mercaptoethanol, 0.4 mL of 10% Nonidet P-40 solution, 50 μL of ampholyte solution (AS) pH 7.0-9.0, 37.5 μL of AS pH 8.0-9.5, 25 μL of AS pH 9.0-11.0, and deionized water to a final volume of 2mL. This solution can be stored at –20°C.

3. Sample Overlay Buffer. This buffer contains 1.92 g of urea, 0.8 mL of 10% Nonidet P-40 solution, 50 μL of AS pH 7.0-9.0, 37.5 μL of AS pH 8.0-9.5, 25 μL of AS pH 9.0-11.0, and 1.55 mL deionized water. This solution can be stored in frozen aliquots. About 400 μL of sample overlay buffer is enough for 8 gel rods.

Electrophoresis Unit

The use of gel rods requires NEPHGE to be carried out in vertical electrophoretic apparatuses. Various sizes of glass tubes can be used with the different electrophoresis units. The protocol described here is for glass tubes 90 mm in length with 5 mm internal diameter, each of them with a capacity for about 2 mL of gel solution.

Method

1. It is convenient to make up the gel rods in lots of eight. Before use, glass tubes should be soaked for 10 min in Kodak Photoflo 600 diluted 1:600, and dried in an oven.

2. Seal the bottom of the glass tubes with several layers of Parafilm, mark them 2.0 cm from the top and set them up in a holder.

3. For 8 rods, 16 ml of gel solution should be prepared by adding 8.84 g urea, 2 mL of 30% acrylamide stock solution, 400 μL of ampholyte solution (AS) pH 7.0-9.0, 300 μL of AS pH 8.0-9.5, 200 μL of AS pH 9.0-11.0, and 3.75 mL deionized water in a vacuum flask. Degas and add 3 mL of 10% Nonidet P-40 solution; dissolve urea by immersing the vacuum flask in warm water, assisting dissolution by moderate swirling. Now add 32 μL of 10% APS and 22.4 μL TEMED. Without delay, swirl the mixture and proceed to the next step.

4. Pour the acrylamide solution into the glass tubes to 2.0 cm from the top using a pipette or a syringe fitted with thin tubing. Rods should be flicked hard when half full to dislodge air bubbles trapped at the bottom. Using a syringe, carefully overlay the acrylamide solution with a small amount of deionized water, cover the tops of the tubes with Parafilm and leave to polymerize for at least 2 h.

5. After polymerization is complete, pour off water overlay, add 80 μL of Lysis Buffer, overlay with a small amount of deionized water, and leave for at least 30 min. Pour off the Lysis Buffer, remove Parafilm from the bottom of the glass tube and replace with dialysis tubing,

being careful to avoid trapping air bubbles in the bottom of the gel tube. Roll a rubber band onto the glass tube to hold dialysis tubing in place.

6. Mount the gel tubes in the electrophoresis apparatus with the Cathode Buffer (20 mM NaOH) in the bottom (lower) reservoir. Load the sample (80μL or more) mixed with Lysis Buffer onto the gel surface using a pipette or a syringe maintained at 37°C, and overlay the sample with 40 μL of Sample Overlay Buffer. Fill the remaining space at the top of the glass tube with Anode Buffer (15 mM orthophosphoric acid, H_3PO_4). Pour the Anode Buffer into the upper reservoir. The gel rods must be fully submerged in the anode solution.

7. Attach the electrophoresis unit to an electric power supply with the positive electrode connected to the upper reservoir. Run gels at 300 V for 4 h.

8. Turn off the power supply, pour off the anode buffer, and remove the gel tubes from the apparatus. Note the orientation of each gel rod. Force each gel rod out of the glass tubes by air pressure from a syringe connected to the rubber tubing and exude them into a sample tube (e.g., test tube) containing 23.75 mL of SDS Equilibration Solution. Add 1.25 mL of β-mercaptoethanol and place in a water bath at 35°C for 2 h before loading onto the second dimension gel. Gels can be stored frozen before or after equilibration.

Second dimension fractionation by SDS-PAGE

Fractionation in the second dimension is SDS-PAGE with a discontinuous buffer system (see section above on SDS-PAGE). For this dimension, a slab electrophoresis apparatus is used. The gels are as for IEF second dimension fractionation, except that 17.5 % acrylamide gels are preferred because this results in less contamination from ampholyte solutions in the lower region of the second dimension gel.

After electrophoresis, turn off the power supply, fix and stain the second dimension gel as described previously (see section above on SDS-PAGE), or use it to establish a western blot (see section below on western blotting).

First dimension IEF and NEPHGE gel rods can be loaded with their tops towards the center of the same second dimension slab gel. In this case, a 17.5% acrylamide gel should be used.

Sample Preparation for First Dimension IEF or NEPHGE Fractionation

The following is a protocol for extraction of storage proteins before fractionation by IEF or NEPHGE.

Sample Buffer Stock. Dissolve 2 g of SDS in 18.3 mL of deionized water and add 10 mL glycerol plus 10 mg of Pyronin Y dye. This Sample Buffer Stock can be kept at room temperature for months.

Sample Buffer. To prepare the Sample Buffer, mix 1.7 mL of the Sample Buffer Stock with 4 mL of deionized water and add 0.3 mL of β-mercaptoethanol.

For IEF fractionation of endosperm proteins, remove embryos from grains before crushing. Weigh 0.05 g of crushed grain into a centrifuge tube. Then add 0.5 ml of sample buffer, mix, and leave at room temperature for 2 h, vortexing occasionally. Transfer to a boiling water bath for 2-3 min, remove, and allow to cool. Centrifuge for 5 min, decant supernatant, and load 20-30 μL of supernatant onto each gel rod.

For NEPHGE fractionation of endosperm proteins, use crushed grains only because milled flour gelatinizes with the Sample Buffer and no supernatant can be withdrawn. Weigh the samples into 1.5 mL microcentrifuge tubes and add the Sample Buffer in the amounts of 0.3 mL /16 mg of sample. Mix and leave at 37°C overnight. Centrifuge for 5 min. Withdraw 50 μL of the supernatant, placing it into a microcentrifuge tube maintained at 37°C, and add to it 50 μL of Lysis Buffer (see First dimension fractionation by NEPHGE, #2). Load immediately.

IEF or NEPHGE Fractionation Using Immobiline Strips

Immobilized pH gradients (IPGs) for IEF were introduced in 1982 (Bjellqvist *et al* 1982), and a basic protocol for two-dimensional electrophoresis IEF × SDS-PAGE with IPGs was established in 1988 (Gorg *et al* 1988). The pH gradient is built up by a small number of defined chemicals covalently grafted to the polyacrylamide matrix, allowing steady-state focusing with high reproducibility. Moreover, basic proteins can be clearly separated under equilibrium conditions using IPGs up to pH 12.

The first dimension IPG isoelectrofocusing is performed in individual IPG strips, which are re-hydrated to 0.5 mm thickness with a solution containing 8M urea, 0.5% Nonidet P-40, and 0.2% β-mercaptoethanol or dithiothreitol. Concentrated protein extracts can be added to this solution.

The amount of sample entry is critical; the best results being obtained using diluted samples dissolved in Lysis Buffer. Many companies, including Amersham Biosciences and Fluka, provide free guides and laboratory manuals of two-dimensional electrophoresis with IPGs.

Recently, analytical IPGs have been used to characterize wheat-grain endosperm proteins by isoelectrofocusing across the pH ranges 4.0-7.0 and 6.0-11.0, followed by SDS-PAGE in the second dimension (Skylas *et al* 2000).

Western Blotting

The technique of protein blotting developed from the work of DNA blotting as originally described by Southern (1975), but instead of using capillary action for transfer, a method of electroelution was devised by Towbin *et al* (1979). This gave much improved transfer and has developed into the standard method. The term "western blotting" was adopted by Burnett (1981) in accordance with the previously used nomenclature for nucleic acid blotting (Southern, 1975) and now covers essentially all forms of protein blotting.

The use of western blotting enables proteins to be stabilized within a membrane to make them more accessible for detection. The proteins can be identified, by probing after electrophoretic separation, using a number of different methods. The procedure is very sensitive in that minor components of a complex mixture can be identified. It is also very convenient in that the membranes, after transfer, can be processed immediately or stored long-term for future use. The immunodetection of proteins bound to a membrane has major applications in plant biochemistry and molecular biology, which include the identification and semi-quantitative determination of foreign proteins expressed in transgenic plants. The method is usually applied to proteins transferred from electrophoretic separations which allows positive identification to be combined with gaining further information about the protein, including M_r, charge, pI, etc. western blotting provides a wide range of options for choice of membrane type, transfer system, and detection system.

General Principles of Western Blotting

There are essentially four methods of protein transfer from polyacrylamide gels to membranes, which are, in terms of increasing efficiency, simple diffusion, capillary action, vacuum blotting, and electroblotting. However, in practice the method most commonly used is electroblotting.

Many blotting systems are available commercially including both wet-tank and semi-dry blotters. The choice of system will depend on numerous factors. Although the semi-dry method does have some advantages including lower buffer volumes and shorter blotting times, the wet-tank method can give better transfer, especially of large proteins including the high molecular weight proteins of gluten.

Proteins can be transferred from most types of gel although the conditions of transfer must be adjusted accordingly. IEF and acid pH (non-SDS) gels can both be transferred to membranes using 0.7% (v/v) acetic acid as transfer buffer, with the proteins migrating towards the cathode. As an alternative, these gels can be equilibrated in SDS-PAGE blotting buffer containing high concentrations (2% w/v) of SDS before transferring as normal SDS gels (the transfer buffer containing 0.01-0.02 % w/v SDS). If 2-D IEF/SDS-PAGE gels are used they can be treated as standard 1-D SDS-PAGE gels, and equilibrated in transfer buffer for between 30 and 40 minutes prior to electroblotting.

Transfer Membranes

Different types of membrane are available for blotting, with the choice being dependent on the protein under study. Nitrocellulose (NC) is the most commonly used, which is available either pure or supported, and in a range of formats and pore sizes. Membrane with a 0.45 μm pore size is commonly used, however low molecular weight proteins (i.e. below 20kDa) require a smaller pore size, with 0.2 μm being recommended. Companies such as Schleicher and Schuell supply pure NC membrane in a number of different pore sizes. Pure nitrocellulose membrane is extremely fragile and will tear and crack easily, so care must be taken to handle it carefully, and always with gloved hands to avoid contaminating the membrane with protein. Nylon membranes are an alternative to NC and are available in charged or uncharged forms. They are more suitable for nucleic acid transfers, are resistant to tearing, and are very pliable. However, because of their higher protein-binding capacity they can give high non-specific binding and, consequently, require increased concentrations of blocking agent. In addition, nylon membranes are incompatible with the most commonly used anionic protein stains such as Amido Black and Coomassie Blue.

The polyvinyldifluoride (PVDF) membranes, such as ProBlott (Applied Biosystems Inc.) have very high binding capacity, and importantly, can be used for automated microsequencing of proteins directly after transfer (only from non-glycine containing systems). They can be used in the same way as NC after initially pre-wetting in 100% methanol for a short time.

Apparatus

Western blotting equipment is available from a large number of suppliers including Bio-Rad and Amersham Biosciences, who produce both wet-tank and semi-dry blotters. Wet-tanks are available in large and small formats, the size being dependent upon the gel size being used, which in turn is dependent upon the proteins under study.

Materials

1. Transfer buffer: 25 mM Tris, 192 mM glycine, and 20% methanol, pH 8.3, with or without 0.02% (w/v) SDS.
2. Nitrocellulose transfer membrane cut to size of gel.
3. Filter papers (Whatman 3MM) cut to the size of the gel.

Electroblotting of Gels

1. Prior to transfer, the transfer membrane must be fully wetted to remove all air pockets. This is best achieved by carefully dipping one edge into a tray of buffer and allowing capillary action to drive air out whilst continuing to lower the membrane into the buffer. Leave fully wetted under buffer until needed. Disposable gloves must be worn at all times when handling the membrane.

2. When PAGE is complete, the gel orientation can be identified by removing one corner. The gel is then equilibrated in transfer buffer for up to 30 min for a 1.5mm-thick gel.
3. A suitable shallow tray is used in which to place the gel holder of the Trans-Blot cell. This is placed cathode (gray) side down and containing sufficient buffer to maintain all further steps just under liquid.
4. A fully wetted Scotch-Brite pad is placed onto the gel holder and a wetted sheet of Whatman 3MM paper placed onto the pad.
5. The equilibrated gel is carefully placed onto the filter paper so as to avoid trapping air pockets.
6. The transfer membrane is carefully held by two sides and laid onto the gel by touching from the center outwards. It is then gently rolled using a rimless test tube (or similar object) to remove any air pockets and ensuring good contact.
7. The sandwich is completed by placing further filter paper(s) over the membrane, and a second Scotch-Brite pad is placed over the filter paper. The gel holder is held firmly so as not to allow slippage of the membrane on the gel, and closed before being placed into the transfer tank with sufficient buffer to cover the blot.

<u>Notes</u>
1. Proteins subjected to SDS-PAGE are transferred as anions, therefore, the membrane is placed on the anodic side of the gel.
2. In order to produce an identical blot from a gel, it is necessary to place the gel reverse-side down.
3. For standard transfers of up to 5 h the buffer is pre-cooled to 4°C and the system chilled during electroblotting. Voltage is held constant at 60V. When gels have been run during the day, with a typical run time of 3-4 h for a large (16x12 cm) gel, it is often very convenient to transfer overnight at 30V.
4. When transfer is complete the membrane is removed and rinsed in Tris-Buffered Saline (TBS). The bound proteins can be visualized immediately or the membrane can be stored, either wet at 4°C for several days or frozen for longer periods.

Membrane Blocking
   To keep non-specific binding to a minimum during the immunodetection of transferred protein, all reactive sites on the filter must first be blocked. This is usually done with protein, often bovine serum albumin (BSA)(up to 5% w/v), fetal calf serum (10% w/v), or gelatin (3% w/v). We have found that one of the least expensive and most efficient blocking agent is a 5% (w/v) suspension of nonfat dried milk powder (Marvel type) in Tris-buffered saline (TBS). Non-ionic detergents such as Tween-20 can also be used, but it has been reported that Tween-20 can give false positive reactions when used alone (Bird *et al* 1988). A comparison of four

commonly used blocking agents showed that defatted milk powder was the most powerful, but also showed the importance of using the correct blocker for any given monoclonal antibody in order to give maximum immunoreactivity (Hauri and Bucher, 1986).

Materials
1. Tris-buffered Saline (TBS): 20 mM Tris/HCl, 500 mM NaCl, pH 7.5.
2. Blocking Agent: 20 mM Tris/HCl, 500 mM NaCl, pH 7.5, 5% (w/v) nonfat dried milk powder (Marvel type).

Method
To block the membrane, it is washed briefly in TBS before carefully being immersed in sufficient blocking solution to cover the membrane. Shake gently for 1 h at room temperature.

Notes
It has been demonstrated that the blocking should be maintained during immunodetection. The addition of low levels of blocking agent (routinely 1% (w/v) BSA) to the wash solutions and to the solutions containing primary and secondary antibodies has been shown to be beneficial as over 60% of blocking protein is reported to be lost from the membrane during washing steps (Ono and Tuan, 1990).

Detection Methods
A range of methods can be used for detection, including general staining methods for total proteins, or specific detection systems for individual proteins contained within a complex mixture.

Total Protein Stains
The detection of total proteins in NC and PVDF membranes can be made using a variety of anionic dyes, which are also used to stain proteins in polyacrylamide gels. To determine the efficiency of transfer, staining can be done in combination with immunodetection (see section below). The most commonly used stains are CBB-R 250 (Burnette, 1981) which is very suitable for PVDF membrane but gives high background with NC, Amido black (Towbin *et al* 1979), Fastgreen (Reinhart and Malamud, 1982), India ink (Hancock *et al* 1983), and Aurodye (Moeremans *et al* 1985), the latter being sensitive in the low nanogram range.

It is also possible to identify the positions of molecular weight marker proteins using pre-stained markers such as those supplied by Amersham or Bio-Rad. These are clearly visible on the gel whilst the gel is running and also on the membrane after transfer. However, the mobilities of the pre-stained proteins may not correspond precisely to those of unstained markers. Biotinylated marker proteins, which are available from Bio-Rad, can provide an alternative method and have the added advantage of allowing processing as for a normal immunoblot. Avidin conjugated with horseradish peroxidase (avidin-HRP) or alkaline phosphatase (avidin-AP) can be added to the conjugated second antibody prior to incubation. The second antibody will detect the immobilized protein (antigen), the labeled avidin the biotinylated standards.

When staining for total protein it is often more useful to use a reversible staining dye such as Amido black (Harper *et al* 1986) or especially the red dye Ponceau S (Salinovich and Montelaro, 1986). Although not as sensitive as other stains, Ponceau S is very rapid and simple to use. However, before using a reversible stain prior to immunodetection, any possible interference with the immunoreactivity of proteins should be carefully checked.

Protocol for Total Protein Staining with Ponceau S

Materials
The stain is available in both powder and solution form from a number of suppliers, including a 0.1% (w/v) Ponceau S solution in 5% (v/v) acetic acid from Sigma.

Method
1. Wash the membrane briefly with water before adding the stain. Leave in contact for several minutes, with shaking, before pouring off. Retain for reuse.
2. Remove background stain by washing the membrane in distilled water. The stain can then be removed by washing twice with Tween-Tris-Buffered Saline (TTBS) before blocking the membrane.

Notes
The stained membrane can be photographed using a green filter. The proteins of interest and standards, as well as individual tracks, can be marked using a soft grade pencil. The membrane can also be stored safely at this point.

Immunodetection Methods
Detection methods are generally indirect and require a specific ligand or antibody to bind to the immobilized protein, followed by the addition of a labeled second antibody or ligand. Probably the most commonly used method is indirect immunological detection using a specific labeled second antibody. These second antibodies are available commercially from a large range of suppliers and offer a wide choice of labels. They are species specific, and will recognize whole primary antibodies of different classes, antibody subtypes or immunoglobulin fragments. They are supplied with radioactive, fluorescent, or enzymic labels, the choice depending on convenience and the degree of sensitivity required. Much of the early work used radiolabels, especially ^{125}I, with detection by autoradiography.

However, enzymic labels are much safer and easier to use, and produce an insoluble reaction product at the binding site. Horseradish peroxidase (HRP) was the first enzyme used for detection, but is not as sensitive as other systems and the insoluble chromogen does fade on exposure to light. A more sensitive system is alkaline phosphatase (AP) which produces a stable end color when used with one of several substrates, the most sensitive being 5-bromo-4-chloro-3-indolyl phosphate (BCIP) and nitro blue tetrazolium (NBT) (Blake *et al* 1984).

Other methods which are available include enhanced chemiluminescence (ECL) (developed by Amersham) in which the primary antibody is detected by HRP-labeled second antibodies. A Luminol substrate is oxidized by HRP to emit light, which is enhanced 1000-fold. Gold-labeled second antibodies, Protein A, or Protein G can also be used, with high sensitivity, which can be increased by silver enhancement (Bio-Rad). Both Protein A and Protein G bind specifically to the Fc region of the antibody, but do not bind equally well to immunoglobulins of different sub-classes or from different species. They are also not as sensitive as species specific antibodies and are less sensitive because only one ligand molecule binds to each antibody.

Alkaline Phosphatase Detection System

Materials
1. Tris-buffered Saline (TBS): 20mM Tris, 500mM NaCl, pH 7.5.
2. Tween-Tris-buffered Saline (TTBS): TBS containing 0.05% v/v Tween 20.
3. Blocking Agent: 5% (w/v) solution of Marvel in TBS.
4. Antibody Buffer: TTBS containing 1% BSA (w/v).
5. Color Development Reagents:
   (A) 50mg/ml solution of nitro blue tetrazolium (NBT) prepared in 70% (v/v) of N, N-dimethylformamide (DMF).
   (B) 25mg/ml solution of 5-bromo-4-chloro-3-indolyl phosphate (BCIP) in 100% DMF.
   (C) Carbonate buffer, (0.1 M $NaHCO_3$, 1.0 mM $MgCl_2$ adjusted to pH 9.8 with NaOH).

Method
1. Wash the membrane briefly in TBS.
2. The membrane is blocked (see Method, above) by immersion in blocking solution and gently shaken for 1 h at room temperature.
3. The blocking solution is discarded, and the membrane washed for 2 × 5 mins with TTBS.
4. The membrane is placed in the dilute antibody solution. Shake for 1-2 h or overnight if more convenient. The dilution factor for the antibody must be worked out empirically, but can vary from ten-fold to several thousand-fold.

5. The primary antibody solution is removed, and can be retained for possible future use. The membrane is washed with TTBS for 2 × 5 min.
6. Add the alkaline phosphatase-labeled second antibody (anti-species), diluted in 1% (w/v) BSA in TTBS for 1-2 hr with shaking. A dilution factor is given by the supplier, but may require adjustment for optimum results.
7. Discard the second antibody and wash the membrane for 2 × 5 min with TTBS followed by 1 × 5 min with TBS to remove residual Tween 20.
8. Prepare the color reagent solution just prior to use by diluting 0.6 ml of each of solutions A and B into 100 ml of solution C and incubate at 37°C prior to use.
9. Add color reagent mixture to the membrane and incubate at 37°C until full color development occurs with minimal background color.
10. Stop the reaction by washing the membrane in distilled water followed by air drying.

Double Staining Method

Stains used to detect total proteins on membranes are often prepared in methanolic solution, which will shrink NC membrane. This often makes it difficult to directly compare the positions of proteins immunologically detected on total protein-stained membranes, especially when comparing two-dimensional gels, which may contain hundreds of individual spots. A very simple way to overcome this problem is to use a double staining method (Ono and Tuan, 1990).

Materials

0.1% (v/v) India ink (or equivalent) in TTBS.

Method

1. The transferred membrane is immunoreacted as described above, using AP-labeled second antibody and NBT/BCIP color reagents to produce an insoluble colored end product.
2. While the membrane is wet, or following rewetting, add sufficient volume of 0.1%(v/v) India ink in TTBS to cover the membrane. Allow sufficient time to develop the blot.
3. Rinse the membrane in water to remove background stain and air dry.

Notes

It is possible to stain with India ink first, followed by immunostaining. However, it is preferable to immunostain first in order to eliminate the possibility that the dye will affect the immunoreactivity of the proteins.

Alternative Methods

New products on the market give the possibility of a detection system without the need to transfer to a membrane with the detection made within the gel itself (Pierce Chemical Company) in a system called Unblot™ In-Gel chemiluminescent detection kit.

Acknowledgements

Long Ashton Research Station receives grant-aided support from the Biotechnology and Biological Sciences Research Council of the United Kingdom.

References

Anderson, L., and Anderson, N.G. 1977. High resolution two-dimensional electrophoresis of human plasma proteins. PNAS, 74:5421-5425.

Bird, C.R., Gearing, A.J.H., and Thorpe, R. 1988. The use of Tween 20 alone as a blocking agent for immunoblotting can cause artifactual results. J. Immunol. Methods. 106:175-179.

Bjellqvist, B., Ek, K., Righetti P.G., Giannazza, E., Gorg, A., Westermeier, R., and Postel, W. 1982. Isoelectric focusing in immobilized pH gradients: Principle, methodology, and some applications. J. Biochem. Biophys. Methods 6:317-339.

Blackshear, P.J. 1984. Systems for Polyacrylamide Gel Electrophoresis. Methods in Enzymology 104:237-255.

Blake, M.S., Johnston, K.H., Russel-Jones, G.J., and Gotschlich, E.C. 1984. A rapid, sensitive method for detection of alkaline phosphatase-conjugated anti-antibody on western blots. Anal. Biochem. 136:175-179.

Bollag, D.M., Rozycki, M.D., and Edelstein, S.J. 1996. Protein Methods, Second Edition. Wiley-Liss: New York.

Borneo, R., and Khan, K. 1999. Glutenin Protein Changes During Breadmaking of Four Spring Wheats: Fractionation by Multistacking SDS Gel Electrophoresis and Quantification with High-Resolution Densitometry. Cereal Chem. 76:718-726.

Brown, J.W.S., and Flavell, R.B. 1981. Fractionation of wheat gliadin and glutenin subunits by two-dimensional electrophoresis and the role of group 6 and group 2 chromosomes in gliadin synthesis. Theor. Appl. Genet. 59:349-359.

Brown, J.W.S., Kemble, R.J., Law, C.N., and Flavell, R.B. 1979. Control of endosperm proteins in *Triticum aestivum* (var. 'Chinese Spring') and *Aegilops umbellata* by homoeologous group 1 chromosomes. Genetics 93:189-200.

Burnette, W.N. 1981. "Western Blotting"; Electrophoretic transfer of proteins from sodium dodecylsulphate-polyacrylamide gels to unmodified nitrocellolose and radiographic detection with antibody and radioiodinated Protein A. Anal. Biochem. 112:195-203.

Bushuk, W., and Zillman, R.R. 1978. Wheat Cultivar Identification by Gliadin Electrophoregrams. I. Apparatus, Method and Nomenclature. Can. J. Plant Sci. 58:505-515.

Cooke, R.J., editor. 1992. Handbook of Variety Testing—Electrophoresis Handbook: Variety Identification. The International Seed Testing Association: Zurich.

Cutting, J.A. 1984. Gel Protein Stains: Phosphoproteins. Methods in Enzymology 104:451-455.

Dupuis, B., Bushuk, W., and Sapirstein, H.D. 1996. Characterization of Acetic Acid Soluble and Insoluble Fractions of Glutenin of Bread Wheat. Cereal Chem. 73:131-135.

Gander, J.E. 1984. Gel Protein Stains: Glycoproteins. Methods in Enzymology 104:447-451.

Garfin, D.E. 1990. One-Dimensional Gel Electrophoresis. Methods in Enzymology 182:425-441.

Gianibelli, M.C., Gupta, R.B., Lafiandra, D., Margiotta, B., and MacRitchie, F. 2001. Polymorphism of High M_r Glutenin Subunits in *Triticum tauschii*: Characterisation by Chromatography and Electrophoretic Methods. J. Cereal Sci. 33:39-52.

Gorg, A., Postel, W., and Gunther, S. 1988. The current status of two-dimensional electrophoresis with immobilized pH gradients. Electrophoresis 9:531-546.

Gupta, R. B., and MacRitchie, F. 1991. A Rapid One-step One-dimensional SDS-PAGE Procedure for Analysis of Subunit Composition of Glutenin in Wheat. J. Cereal Sci. 14:105-109.

Hames, B.D., and Rickwood, D., editors. 1990. Gel Electrophoresis of Proteins—A Practical Approach, Second Edition. IRL Press at Oxford University Press: Oxford.

Hancock, K., and Tsang, V.C.W. 1983. India ink staining of proteins on nitrocellulose paper. Anal. Biochem. 133:157-162.

Harper, D.R., Liu, K.-M., and Kangro, H.O. 1986. The effect of staining on the immunoreactivity of nitrocellulose-bound proteins. Anal. Biochem. 157:270-274.

Hauri, H-P., and Bucher, K. 1986. Immunoblotting with monoclonal antibodies: Importance of the blocking solution. Anal. Biochem. 159:386-389.

Holt, L.M., Austin, R., and Payne, P.I. 1981. Structural and genetical studies on the high-molecular-weight subunits of wheat glutenin. 2. Relative isoelectric points determined by two-dimensional fractionation in polyacrylamide gels. Theor. Appl. Genet. 60:237-243.

Huang, D.Y., and Khan, K. 1996. Unexpected Solubility Changes in Wheat Proteins During Fermentation and Oven Stages of the Breadmaking Process. Cereal Chem. 73:512-513.

Huang, D.Y., and Khan, K. 1997c. Quantitative Determination of High Molecular Weight Glutenin Subunits of Hard Red Spring Wheat by SDS-PAGE. II. Quantitative Effects of Individual Subunits on Breadmaking Quality Characteristics. Cereal Chem. 74:786-790.

Huang, D.Y., and Khan, K. 1997a. Characterization and Quantification of Native Glutenin Aggregates by Multistacking Sodium Dodecyl Sulfate Polyacrylamide Gel Electrophoresis (SDS-PAGE) Procedures. Cereal Chem. 74:229-234.

Huang, D.Y., and Khan, K. 1997b. Quantitative Determination of High Molecular Weight Glutenin Subunits of Hard Red Spring Wheat by SDS-PAGE. I. Quantitative Effects of Total Amounts on Breadmaking Quality Characteristics. Cereal Chem. 74:781-785.

Huang, D.Y., and Khan, K. 1998. A Modified SDS-PAGE Procedure to Separate High Molecular Weight Glutenin Subunits 2 and 2*. J. Cereal Sci. 27:237-239.

Jackson, E.A., Holt, L.M., and Payne, P.I. 1983. Characterisation of high molecular weight gliadin and low-molecular-weight glutenin subunits of wheat endosperm by two-dimensional electrophoresis and the chromosomal localisation of their controlling genes. Theor. Appl. Genet. 66:29-37.

Jones, R. W.; Taylor, N. W., and Senti, F. R. Electrophoresis and Fractionation Of Wheat Gluten. Arch. Biochem. Biophys. 1959; 84:363-376.

Khan, K., and Huckle, L. 1992. Use of Multistacking Gels in Sodium Dodecyl Sulfate-Polyacrylamide Gel Disaggregation of the Glutenin Protein Fraction. Cereal Chem. 69:686-688.

Khan, K., Hamada, A.S., and Patek, J. 1985. Polyacrylamide Gel Electrophoresis for Wheat variety Identification: Effect of Variables on Gel Properties. Cereal Chem. 62:310-313.

Khan, K., McDonald, C.E., and Banasik, O.J. 1983. Polyacrylamide Gel Electrophoresis of Gliadin Proteins for Wheat Variety Identification— Procedural Modification and Observations. Cereal Chem. 60:178-181.

Khan, K. 1982. Polyacrylamide Gel Electrophoresis of Wheat Gluten Proteins. Bakers Digest 56:14-19.

Khan, K., and Bushuk, W. 1977. Studies of Glutenin. IX. Subunit Composition by Sodium Dodecyl Sulfate-Polyacrylamide Gel Electrophoresis at pH 7.3 and 8.9. Cereal Chem. 54:588-596.

Kim, H.R., and Bushuk, W. 1995. Salt Sensitivity of Acetic Acid-Extractable Proteins of Wheat Flour. J. Cereal Sci. 21:241-250.

Laemmli, U.K. 1970. Cleavage of Structural Proteins during the Assembly of the Head of Bacteriophage T4. Nature 227:680-685.

Lafiandra, D., Turchetta, T., D'Ovidio, R., Anderson, O. D., Facchiano, A. M., and Colonna, G. Conformational polymorphism of high M-r glutenin subunits detected by transverse urea gradient gel electrophoresis. J. Cereal Sci. 1999; 30:97-104.

Lookhart, G.L., Jones, B.L., Hall, S.B., and Finney, K.F. 1982. An Improved Method for Standardizing Polyacrylamide Gel Electrophoresis of Wheat Gliadin Proteins. Cereal Chem. 59:178-181.

Lukow, O.M., Payne, P.I., and Tkachuk, R. 1989. The HMW Glutenin Subunit Composition of Canadian Wheat Cultivars and their Association with Bread-Making Quality. J. Sci. Food Agric. 46:451-460.

Mackie, A.M., Lagudah, E.S., Sharp, P.J., and Lafiandra, D. 1996. Molecular and Biochemical Characterisation of HMW Glutenin Subunits from *T. tauschii* and the D Genome of Hexaploid Wheat. J. Cereal Sci. 23:213-225.

Merril, C.R., Goldman, D., and VanKeuren, M.L. 1984. Gel Protein Stains: Silver Stain. Methods in Enzymology 104:441-447.

Merril, C.R. 1990. Gel-Staining Techniques. Methods in Enzymology 182:477-488.

Moeremans, M., Daneels, G., and De Mey, J. 1985. Sensitive colloidal metal (gold and silver) staining of protein blots on nitrocellulose membranes. Anal. Biochem. 145:315-321.

Neuhoff, V., Stamm, R., Pardowitz, I., Arold, N., Ehrhardt, W., and Taube, D. 1990. Essential problems in quantification of proteins following colloidal staining with Coomassie Brilliant Blue dyes in polyacrylamide gels, and their solution. Electrophoresis 11:101-117.

Neuhoff, V., Arold, N., Taube, D., and Ehrhardt, W. 1988. Improved staining of proteins in polyacrylamide gels including isoelectric focusing gels with clear background at nanogram sensitivity using Coomassie Brilliant Blue G-250 and R-250. Electrophoresis 9:255-262.

Ng, P.K.W., and Bushuk, W. 1987. Glutenin of Marquis Wheat as a Reference for Estimating Molecular Weights of Glutenin Subunits by Sodium Dodecyl Sulfate-Polyacrylamide Gel Electrophoresis. Cereal Chem. 64:324-327.

O'Farrell, P.H. 1975. High resolution two dimensional electrophoresis of Proteins. J. Biol. Chem. 250:4007-4021.

O'Farrell, P.Z., Goodman, N.M., and O'Farrell, P.H. 1977. High resolution two-dimensional electrophoresis of basic as well as acid proteins. Cell 12:1133-1142.

Ono, T., and Tuan, R.S. 1990. Double staining of immunoblot using enzyme histochemistry and India ink. Anal. Biochem. 187:324-327.

Patton, W.F. 2001. Detecting proteins in polyacrylamide gels and on electroblot membranes. Pages 65-86 in: Proteomics—From protein sequence to function. S.R. Pennington and M.J. Dunn, eds. BIOS Scientific Publishers Ltd.: Oxford.

Reinhart, M.P., and Malamud, D. 1982. Protein transfer from isoelectric focusing gels: The native blot. Anal. Biochem. 123:229-235.

Reisner, A.H. 1984. Gel Protein Stains: A Rapid Procedure. Methods in Enzymology 104:439-441.

Rogers, W.J., Payne, P.I., and Harinder, K. 1989. The HMW Glutenin Subunit and Gliadin Compositions of German-Grown Wheat Varieties and their Relationship with Bread-Making Quality. Plant Breeding 103:89-100.

Rosenberg, I.M. 1996. Protein Analysis and Purification—Benchtop Techniques. Birkhauser: Boston.

Salinovich, O., and Montelaro, R.C. 1986. Reversible staining and peptide mapping of proteins transferred to nitrocellulose after separation by sodium dodecylsulphate-polyacrylamide gel electrophoresis. Anal. Biochem. 156:341-347.

Skylas, D.J., Mackintosh, J.A., Cordwell, S.J., Basseal, D.J., Walsh, B.J., Harry, J., Blumenthal, C., Copeland, L., Wrigley, C.W., and Rathmell, W. 2000. Proteome approach to the characterisation of protein composition in the developing and mature wheat-grain endosperm. J. Cereal Sci., 32:169-188.

Southern, E.M. 1975. Detection of specific sequences among DNA fragments separated by gel electrophoresis. J. Mol. Biol. 98:503-517.

Towbin, H., Staehelin, T., and Gordon, J. 1979. Electrophoretic transfer of proteins from polyacrylamide gels to nitrocellulose: Procedure and some applications. Proc. Natl. Acad. Sci. USA. 76:4350-4354.

Wilkins, M.R., Williams, K.L., Appel, R.D., and Hochstrasser, D.F. 1997. p. 24 in Proteome Research: New Frontiers in Functional Genomics. Springer: Berlin.

Wilson, C.M. 1983. Staining of Proteins on Gels: Comparison of Dyes and Procedures. Methods in Enzymology 91:236-247.

Woychik, J. H.; Boundy, J.A., and Dimler, R. J. Starch Gel Electrophoresis Of Wheat Gluten Proteins With Concentrated Urea. Arch. Biochem. Biophys. 1961; 94:477-482.

Zhen, Z., and Mares, D. 1992. A Simple Extraction and One-step SDS-PAGE System for Separating HMW and LMW Glutenin Subunits of Wheat and High Molecular Weight Proteins of Rye. J. Cereal Sci. 15:63-78.

Zhu, J., and Khan, K. 2001. Effects of Genotype and Environment on Glutenin Polymers and Breadmaking Quality. Cereal Chem. 78:125-130.

Zhu, J., Khan, K., Huang, S., and O'Brien, L. 1999. Allelic Variation at Glu-D1 Locus for High Molecular Weight (HMW) Glutenin Subunits: Quantification by Multistacking SDS-PAGE of Wheat Grown Under Nitrogen Fertilization. Cereal Chem. 76:915-919.

Zhu, J., and Khan, K. 1999. Characterization of Monomeric and Glutenin Polymeric Proteins of Hard Red Spring Wheats During Grain Development by Multistacking SDS-PAGE and Capillary Zone Electrophoresis. Cereal Chem. 76:261-269.

Chapter 4

HPLC of Gluten Monomeric Proteins

George L. Lookhart
USDA-ARS, Grain Marketing and Production Research Center,
Manhattan, KS

Scott R. Bean
USDA-ARS, Grain Marketing and Production Research Center,
Manhattan, KS

Jerold A. Bietz (retired)
USDA-ARS, National Center for Agricultural Utilization Research, Peoria, IL

Introduction

High performance liquid chromatography (HPLC) uses a liquid pumping system to accurately deliver solvents through a column packed with 1.5- to 10-μm particles with specific bonded phases. The end result is the ability to separate complex mixtures in minutes. HPLC is a superb tool. It is complementary and often superior to previous methods for characterization of complex cereal proteins. Hundreds of reports describe such studies; these can only be summarized here, with references to more comprehensive reviews. This chapter describes the separation of gluten monomers via two modes of HPLC, reversed-phase and ion-exchange. The monomeric gluten proteins were described by Shewry in the Introduction to this book. In this chapter, we have chosen to use a liberal definition of the term, and also describe some separations of glutenin subunits as monomers.

While many "good" protocols exist; there is no general agreement as to which is best, which implies that this method is not completely mature. First time users *must* become more familiar with the method than can be done by simply reading the present chapter, they must consult some of the original references and definitive reviews cited here. This chapter was taken in part from a broad review on HPLC of cereal proteins by Bietz (2002).

U.S. Department of Agriculture, Agricultural Research Service, Northern Plains Area, is an equal opportunity/affirmative action employer and all agency services are available without discrimination.

Mention of firm names or trade products does not constitute endorsement by the U.S. Department of Agriculture over other not mentioned.

RP-HPLC

Principles

RP-HPLC was so named because the chemical functionality (hydrocarbon - non-polar) of column bonded phases has opposite polarity to "normal" phase (Si-OH or polar) columns. RP-HPLC is the most often used HPLC mode for gluten protein analysis. Its resolution equals or exceeds that of most other methods. It is also fast, reproducible, sensitive, quantifiable, and gives good recovery. Most importantly, however, it complements other methods since it fractionates proteins on the basis of different surface hydrophobicities. Selection of optimal RP-HPLC columns requires consideration of many factors, including support type, hydrophobic ligand, pore size, particle size, column dimensions, and silanization (Neville 1996, Aguilar and Hearn 1996).

Reagents

All reagents for HPLC analyses should be of highest possible purity, typically HPLC grade. Impurities in reagents used in mobile phases can cause UV absorption or directly interact with proteins. Generally, "HPLC grade" solvents are suitable, but spectral-grade solvents may not be. Water quality is especially important. Bottled HPLC-grade water may deteriorate with time; the best water (especially for RP-HPLC) is produced by freshly processing distilled water in the laboratory with a device having mixed-bed ion exchange, carbon (organic), and particulate (0.45 µm) filters. All solvents should be deaerated by vacuum, filtration (0.2-0.5 µm), helium sparging, and/or sonication. A flat baseline is the ultimate test of solvent suitability; the presence of peaks or an increasing baseline in blank runs indicates solvent problems.

Apparatus

Several companies produce HPLC instruments with a wide variety of configurations. Important features to consider are; pumping system, autosampler, detectors, column heater, and ease of using both the system and software. A simple schematic is shown in Fig. 1. The pumping system is the most important as it must be able to accurately produce constant flow rates at back pressures of up to 5000 psi and gradients or blends of organic and polar solvents.

Procedure

Initial RP-HPLC separations of gluten proteins, on 250 × 4.1 mm i.d. columns packed with 300 A C_{18}-coated silica 5- to 10-µm particles, showed the the excellent stability and reproducibility of wide-pore columns (Bietz 1983).

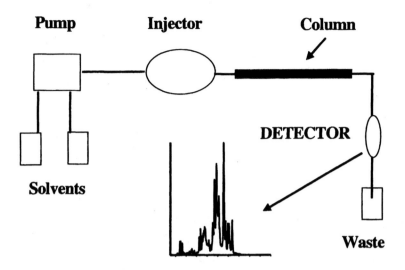

Fig. 1. HPLC schematic; showing a high pressure liquid pump with two solvents connected, an injector system for placing samples into the pressurized liquid line, a column with a specified bonded phase for separating the materials, a detector for indicating when compounds are eluting, and a waste container for collecting the effluent. The results of the detector changes are shown in the pattern profile.

Many additional wide-pore "protein" columns having C_{18}, C_8, C_4, C_3, cyano, diphenyl, and cyanopropyl bonded phases were tested with wheat gliadins and glutenin subunits (Burnouf and Bietz 1984a, Bietz *et al* 1983, Kruger and Marchylo 1984). Some differences in resolution and selectivity were apparent. Separations achieved on many commercial C_8 and C_{18} columns of comparable bonded phases were preferred. Columns from different sources, having comparable bonded phases, gave similar separations. Column stability and age affected reproducibility and resolution, and columns of the same type varied among, or even within, manufacturers (Marchylo 1994). Purchasing multiple columns of the same lot is desirable.

Shorter RP-HPLC columns combined with higher flow rates, steeper gradients, or elevated temperature are often acceptable (Kruger and Marchylo 1984, Marchylo *et al* 1988, Bietz and Cobb 1985). Their resolution can equal that of longer columns since most proteins either bind tightly or not at all to hydrophobic ligands as mobile phase conditions change (Marchylo 1994). For example, Huebner and Bietz (1995) used 1.2-3.2 mm × 15 cm C_4 columns to fractionate gliadins and glutenin subunits and to differentiate cultivars. Flow rate, sample size, analysis time, solvent consumption, and cost per analysis were reduced greatly, while providing excellent resolution. Some resolution may be sacrificed for varietal identification. Lookhart (1992) reported the use of 5-μm particle C_{18} Vydac guard columns (3.2mm × 1 cm) to fractionate gliadins for cultivar identification in less than 3 min (unpublished data). However, column lifetime was short; separations deteriorated after 75-80 analyses. Further progress in developing rapid RP-HPLC methods for cereal proteins is undoubtedly possible: Chen and Horvath (1995) showed that 1.5-2 μm porous support particles provide rapid analyses of proteins, and that novel micropellicular 4000 A pore size supports, due to superior mass transfer, may be highly suitable for rapid and efficient HPLC of biological macromolecules.

Another important development in optimizing cereal protein RP-HPLC was reported by Marchylo *et al* (1992a). Sterically protected wide-pore monofunctional-silane bonded C_8 and CN columns (Zorbax R-300) gave better resolution and reproducibility of gliadins and glutenin subunits than did conventional silica-based columns, and were more stable. A typical separation is shown in Fig. 2.

Procedures

Solvent uniformity is essential for HPLC reproducibility, especially with gradient elution. In early studies (7), equipment limitations often precluded use of pure water and acetonitrile (ACN) (+ 0.1% of an ion-pairing agent, trifluoroacetic acid [TFA]) to generate gradients because of solvent outgassing. Instead, 15 and 80% aqueous ACN (+ 0.1% TFA) were used since all proteins eluted in this hydrophobicity range and outgassing was

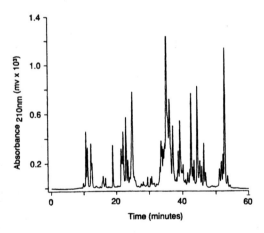

Fig. 2 RP-HPLC on a Zorbax Rx-300-C8 column of total gluten proteins (gliadins and glutenins) extracted under reducing conditions from the wheat c.v. Neepawa. (From Marchylo *et al* 1992a, with permission).

less severe. With such solvents, however, solute elution positions may change slightly with time due to evaporation of solvent components (Bietz and Cobb 1985). Mixed ACN/water/TFA solvents can also be difficult to prepare reproducibly. Outgassing during gradient formation is usually less severe with modern HPLC systems. Consequently, pure water and acetonitrile (+ TFA) can generally be used, especially if maintained under helium and / or the detector flow cell is slightly pressurized.

Early (ca. 1980) RP-HPLC separations of non-cereal proteins on wide pore columns achieved excellent resolution using the ion-pairing agent TFA with aqueous gradients of ACN, propan-1-ol / propan-2-ol, or other organic solvents (Mahoney and Hermodson 1980, Hearn 1980). Comparable conditions were applied to cereal proteins by Bietz (1983). ACN gradients with ca. 0.1% TFA generally gave the best results. ACN has low viscosity, is relatively transparent at ca. 210 nm, and generally provides better resolution than gradients of propan-1-ol or propan-2-ol or methanol. TFA has relatively high absorbance at 210 nm, but very low absorbance at ca. 0.1%. Higher TFA concentrations may modify selectivity (Bietz 1983), and inclusion of 0.5% TFA in RP-HPLC solvents enhances resolution of some cereal proteins (Lookhart, unpublished results). Lower TFA concentrations (e.g., 0.05%) are often used to prevent deamidation (Huebner and Bietz 1984). Sloping baselines can also be controlled by varying solvent TFA concentrations (Huebner and Bietz 1987). Other ion-pairing agents (including heptafluorobutyric acid and sodium dodecyl sulfate [SDS]) (Bietz 1983) may modify selectivity, but have not been widely used. Similarly, solvents near pH neutrality, though little used, may provide excellent separations. Vensel et al reported superior separations of γ-gliadins with an acetonitrile:propanol-2-ol (3:1) solvent system containing 0.02 M or 0.05 M hexafluoroacetone and adjusted to pH 7.2 with ammonia (Vensel et al 1989).

The use of SDS in RP-HPLC solvents has been reported (Bietz 1983). This is somewhat surprising, since SDS binds strongly to proteins and may mask hydrophobic surfaces; nevertheless, SDS solvents can modify selectivity and enhance resolution in RP-HPLC, and SDS is a powerful solvating agent. Further studies are required to explain and optimize the use of SDS in RP-HPLC. High concentrations of urea (5-6 M) and reducing agents, such as dithiothreitol, can also be incorporated into RP-HPLC solvents to ensure protein solubility, modify selectivity, and give excellent separations (Seilmeier et al 1987, Seilmeier et al 1991a).

Column temperature is very important in optimizing HPLC separations, especially in RP-HPLC. Early RP-HPLC protein separations were often done at ambient temperature. Fluctuations in "room" temperature, however, significantly decreased run-to-run reproducibility of elution times (Bietz and Cobb 1985). Thus, maintaining constant column temperature (and, ideally, preheating solvents to this temperature) is essential for good reproducibility.

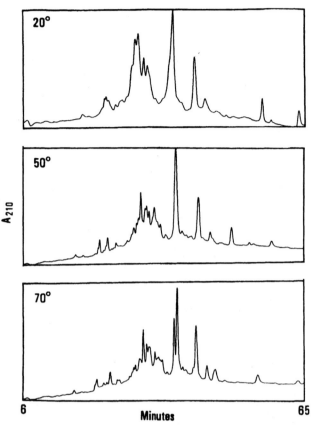

Fig. 3 RP-HPLC of gliadins from wheat cv. Anza at 20, 50 and 70°C. (Taken from Bietz and Cobb 1985, with permission).

Similarly, early RP-HPLC studies of proteins suggested that resolution generally did not increase with temperature, so a maximum of ca. 50°C was generally used. Bietz and Cobb (1985), however, found marked improvement in resolution of gliadins between 50 and 70°C (Fig. 3). Variation in resolution and selectivity with temperature was shown in subsequent studies (Marchylo *et al* 1988, Chloupek *et al* 1994), with higher temperature separations generally giving better results. In an extreme example by Chen and Horvath (1995) four standard proteins were resolved in less than 10 sec at 120°C! Sometimes, however, as for purothionins, lower temperatures enhance separation (Jones *et al* 1985). The general importance of temperature in RP-HPLC, and the possibility of improving, or fine-tuning, separations by varying temperature has been recognized

(Ooms 1996). There are no standard procedures recognized. Each laboratory has their own set of conditions that are generally similar

General Procedure
The following procedure is recommended for gliadins.
1. Column -
    Use a C_4, C_8, or C_{18} bonded phase column, the latter two are most similar but the C_{18} column is more often used.
2. Sample preparation -
    Extract 100 mg sample (flour or finely ground grain) with 1 ml 50% propan-1-ol in a 2 ml micro centrifuge tube by continual vortexing on a VortexGenie2 for 5 min and then centrifuging for 2 min at 14,000 × G. (Smaller samples can be used. As little as 10 mg can be extracted with 100 µL of solvent can be used - ½ kernel or less.) Transfer the liquid carefully into an HPLC autosampler vial. If there are problems in the separation due to the extraction solvent, the extracts can be dried by vacuum and redissolved in the HPLC starting buffer.
3. Loading -
    Inject 5-50 µL (usually 5 - 10 µL) into the HPLC system. Use larger volumes if you want to collect fractions separated by the column.
4. Pumping conditions -
    Elute at 70°C and 1 ml/min with a multi-step gradient starting at 25% B, to 35% B at 3 min, to 53% B at 25 min, to 75% B at 26 min., and staying at 75% B for 3 min and then returning to initial conditions, 25% B. Equilibrate with initial concentrations for 10 min. (Longer gradients (time, %B) can be used.)
    Solvent A: 0.1 % (wt/v) (667µL/L) TFA in water.Solvent B: 0.1 % (wt/v) (667µL/L) TFA in acetonitrile.
5. Detection -
    Record absorbance at 200, 210, 220, 254, and/or at 280 nm, (210 nm is preferred.)
6. Fractions -
    Gliadins are eluted in the following order; omega, alpha-beta, and finally gamma (Lookhart and Albers 1988)

Recommended Glutenin Procedure
The recommended glutenin procedure is nearly identical to the gliadin procedure. The only real difference is the extraction conditions, where the gliadins are extracted first and discarded (in order to get cleaner glutenin patterns). The glutenins are then extracted with 50% propan-1-ol containing 1% DTT (dithiothreitol), vortexed for 5 min, centrifuged, and the liquid pipetted into autosampler vials. The HPLC column, solvents, gradients and temperatures are identical to those used for gliadins. The HMW glutenins are eluted first followed by the LMW subunits.

Applications

Early studies showed resolution of wheat proteins by RP-HPLC to be equal or superior to that given by other methods (Bietz 1983). For example, gliadin was resolved into more than fifty peaks and shoulders in 50 min on a C_{18} column at 60°C (Bietz and Cobb 1985). RP-HPLC is also more rapid, sensitive, and reproducible than size-exclusion HPLC. RP-HPLC gives good quantitation and recovery; and it is suitable for both preparative and analytical separations. In RP-HPLC, proteins resolve primarily based on differences in surface hydrophobicity, which complements techniques such as gel electrophoresis that are based on size or charge (Bietz 1983, Lookhart and Albers 1988).

RP-HPLC has been used to study accumulation of proteins during wheat development (Gupta *et al* 1996, Huebner *et al* 1990). Muller and Wieser (1995) fractionated gliadins by RP-HPLC, digested them enzymatically, and characterized intra-molecular disulfide cross-links by RP-HPLC before and after reduction. Popineau and Pineau (1987) separated gliadins on the basis of surface hydrophobicity using RP-HPLC, hydrophobic interaction chromatography, and non polar ligand binding. The methods gave similar information, but RP-HPLC provided higher resolution. RP-HPLC can be used to relate wheat protein hydrophobicity to structures, aggregation tendencies, and functional properties, such as viscoelasticity, emulsification, and foaming (Popineau 1994). Kruger *et al* (1988) used RP-HPLC to isolate and characterize sequentially extracted wheat proteins, relating the amounts to dough strength. Lookhart *et al* (1989) used RP-HPLC to compare gliadins from flour mill streams and fractions. Bean and Lookhart (1997) used a two-dimensional method, which combined RP-HPLC with capillary electrophoresis to characterize gliadins and glutenins. Surface contour plots constructed from these procedures show superior resolution, differentiate genotypes, and reveal quality-related proteins. Wieser *et al* (1994) reviewed preparative and analytical RP-HPLC of glutenin subunits, showing how protein compositions relate to technological and rheological properties.

RP-HPLC is especially useful to characterize the numerous component polypeptides, or subunits, of polymeric wheat glutenin. In initial studies (Bietz *et al* 1984a), 15-20 major subunits were resolved; early peaks contained less-hydrophobic high MW subunits, and later peaks contained more hydrophobic lower-MW subunits. In a series of papers over several years, Wieser *et al* (1989a, b), Seilmeier *et al* (1987, 1991a) and Werbeck *et al* (1989) significantly improved glutenin RP-HPLC methods. Glutenin was reduced, extracted with water/alcohol mixtures, and fractions were characterized by RP-HPLC. Other studies recommended fractionation of glutenin by 70% aqueous ethanol under reducing conditions before RP-HPLC. Low MW glutenin subunits are ethanol-soluble, and ethanol-insoluble subunits are primarily high MW. Each variety exhibited an HPLC pattern specific for a set of soluble subunits, and their high MW

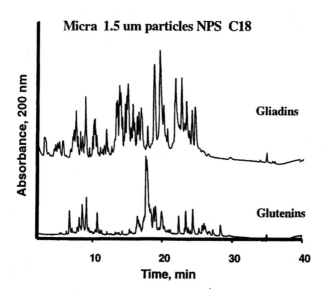

Fig. 4. RP-HPLC patterns of gliadin and reduced but non-alkylated glutenin fractions from wheat cv. Karl on Micra, 1.5 μm non-porous silica C_{18} column. The HMW-GS are on the left (5-12 min) the LMW are on the right (15-30 min). (Taken from Lookhart 1997, with permission.)

subunit compositions related to baking quality. Several solvent systems were developed to characterize glutenin subunits, and used to relate qualitative and quantitative compositions to genotype and quality. RP-HPLC patterns of gliadin and non-alkylated reduced glutenin extracts of flour from wheat cv Karl separated on Micra non-porous 1.5 μm silica media are shown in Fig. 4 (Lookhart 1997). The high resolution gliadin RP-HPLC pattern shows over 50 peaks. The glutenin pattern shows that HMW-GS were well resolved in the 5 to 12 min range and LMW-GS in the 15 to 30 min range.

These and similar procedures have significantly enhanced our knowledge of the nature and role of high- and low-MW subunits of glutenin. High-MW subunits differentiate alleles associated with breadmaking quality. For example, Sutton (1991) showed that RP-HPLC could differentiate allelic high MW glutenin subunits in New Zealand cultivars, and that their composition related to breadmaking quality. Margiotta et al (1993) characterized allelic variation among high MW durum glutenin subunits by RP-HPLC, and showed the complementary nature of RP-HPLC to other techniques. RP-HPLC can also identify and characterize novel high MW wheat glutenin subunits (Margiotta *et al* 1996). Lafiandra *et al* (1993) used RP-HPLC to show that high-MW glutenin subunits having the same SDS-PAGE (sodium dodecyl sulfate polyacrylamide gel electrophoresis) mobilities, from varieties differing in breadmaking quality, are sometimes not identical, emphasizing the importance of the use of multiple and complementary analytical methods. Weegels *et al* (1995) used RP-HPLC to characterize glutenin isolated by various procedures. Relative amounts and types of high-MW subunits differed between preparations, suggesting that x- and y-type subunits contribute differently to the size and composition of glutenin polymers.

Similarly, RP-HPLC is invaluable for studying low MW subunits of glutenin. Gradients have been optimized to give excellent resolution of these proteins (Wieser *et al* 1989a,b, Seilmeier *et al* 1987). Lew *et al* (1992) purified low-MW glutenin subunits by RP-HPLC and related their sequences to polymer structure. When Masci *et al* (1995) used RP-HPLC to isolate low MW glutenin subunits from durum biotypes differing at the Gli-B1/Glu-B3 loci for sequence analyses, they found that differences in the structure and amount of these proteins are associated with differences in glutenin polymerization and quality. RP-HPLC is also valuable for analyzing minor wheat proteins, including purothionins (Jones *et al* 1985) and lipid-binding proteins (Zawistowska *et al* 1986).

Varietal Identification

HPLC is applicable to selection, identification, registration, and marketing of all plant genotypes. One of the most important applications of HPLC is varietal identification. This is possible because storage protein expression is nearly invariant within genotypes, and "fingerprints" vary little with environment (Kruger and Marchylo 1984, Lookhart and Pomeranz 1985, Cressey 1987, Huebner and Bietz 1988, Marchylo *et al* 1990, Huebner and Gaines 1992, Huebner *et al* 1995) or even upon germination (Kruger and Marchylo 1985). Such analyses may be important during marketing, for quality prediction, or to register varieties, especially as identity preservation becomes more widespread. In addition, selection of known and desired genotypes is increasingly important during breeding and in genetic studies. While other procedures - especially gel electrophoresis - have been widely used for these purposes, the many advantages of RP-HPLC (speed, relative simplicity, automation, and quantification) can easily make it the method of choice. Reviews of this topic are provided by Bietz and Huebner (1994) and Lookhart *et al* (1994).

RP-HPLC of high- or low-MW glutenin subunits may be used to differentiate some wheat genotypes (Burnouf and Bietz 1984a, Seilmeier *et al* 1988, 1991a,b, Wieser *et al* 1989a). However, most HPLC methods for wheat varietal identification use RP-HPLC to analyze gliadins, since these methods are relatively mature, and gliadins, qualitatively the most invariant

70

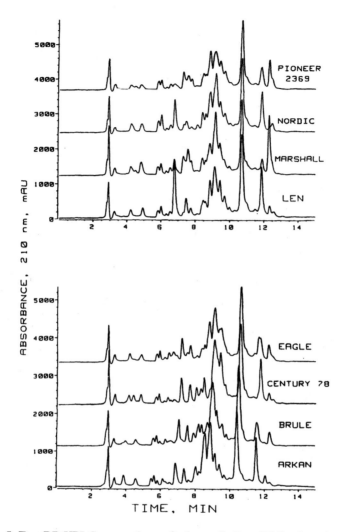

Fig. 5. Fast RP-HPLC separations of wheat gliadins (70% ethanol extracts) for varietal identification of U.S. hard red winter wheats (bottom) and U.S. hard red spring wheats (top). Extracts (10 uL) were analyzed on a Vydac C18 column by elution with a water-acetonitrile gradient at 45°C. (Taken from Lookhart *et al* 1993a, with permission).

wheat storage proteins, are easy to isolate. Bietz (1983) first developed and Bietz and co-workers subsequently optimized procedures to identify wheat varieties by RP-HPLC of gliadins (Burnouf *et al* 1983, Bietz *et al* 1984a,b, Bietz and Cobb 1985, Lookhart and Bietz 1990,). Since then, many workers have used gliadin RP-HPLC to differentiate cultivars (Lookhart *et al* 1985, 1986, Lookhart and Pomeranz 1985, Hansen and Prentice 1985, Wieser *et al* 1987, Cressey 1987, Marchylo *et al* 1988, Ng *et al* 1989, Hay and Sutton 1990, Ram *et al* 1995). Typical RP-HPLC gliadin patterns of 8 U.S. hard red wheats (4 winters [bottom] and 4 springs [top]) are shown in Fig. 5 (Lookhart *et al* 1993a). RP-HPLC for varietal identification was recently reviewed (Barnwell *et al* 1994).

Procedures for computer-assisted data handling and statistical analyses have automated the method and improved its accuracy (Sapirstein *et al* 1989, Scanlon *et al* 1989a,b, Lookhart *et al* 1993a). RP-HPLC analyses of gliadins can be very rapid (Fig. 5), paving the way for inexpensive and routine RP-HPLC varietal identification, and for related applications. RP-HPLC of gliadins also distinguishes biotypes within varieties (Bietz and Cobb 1985, Bietz *et al* 1984a, Lookhart *et al* 1986); knowledge of such genetic heterogeneity is essential for selection during breeding. Similarly, gliadin RP-HPLC can determine the purity of hybrid wheats (McCarthy *et al* 1990). RP-HPLC can determine the proportion of wheat and rye in

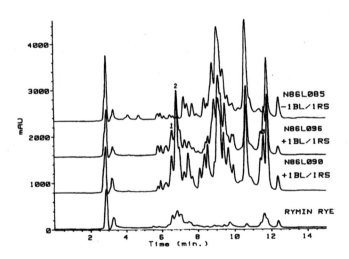

Fig. 6. Differentiation of wheat-rye translocation lines by RP-HPLC patterns of 70% ethanol extracts of wheat (N86L085), wheat-rye translocation lines (N86L090 & N86L096) and rye (cv. Rymin). Peaks numbered 1,2, & 3 are probable rye secalins. (Taken from Lookhart *et al* 1991, with permission.)

mixtures, based on quantification of late-eluting rye secalins (Sutton and Hay 1990), and can identify wheat-rye1BL.1RS translocation lines (Fig. 6) (Lookhart *et al* 1991). Adulteration of durum wheat flour with common wheat can also be identified and quantified by RP-HPLC of gliadins or albumins that consistently differentiate tetraploid and hexaploid genotypes (McCarthy *et al* 1990b, Barnwell *et al* 1994, DeNoni *et al* 1994).

Preparative RP-HPLC

From the outset, it was apparent that RP-HPLC was a valuable preparative method for cereal proteins. Up to 5 mg samples (potentially) can be fractionated on analytical columns with good resolution (Bietz 1983), and this can be scaled up considerably. Electrophoresis of the resulting fractions is a complementary technique that confirms the resolution of RP-HPLC and further fractionates on the basis of size or charge (Bietz 1983, Lookhart and Albers 1988, Bean and Lookhart 1997).

In early studies, Burnouf and Bietz (1984b) used preparative RP-HPLC to isolate durum wheat gliadins associated with pasta quality. Preparative RP-HPLC also fractionated high MW glutenin subunits related to breadmaking quality (Burnouf and Bietz 1984a). To isolate proteins from complex mixtures, however, it is often desirable to combine RP-HPLC with ion-exchange or size-exclusion methods (Huebner and Bietz 1984). Wieser and Belitz (1992), for example, describe a preparative procedure combining RP-HPLC with gel permeation chromatography to isolate celiac-active gliadin peptides.

Others have since used preparative RP-HPLC to purify gliadins for characterization and structural studies, and to relate them to rheological and breadmaking properties (Aibara and Morita 1988, Freedman *et al* 1988, Khan *et al* 1992). Similarly, high- and low-MW glutenin subunits have been fractionated by preparative or semi-preparative RP-HPLC (Wieser *et al* 1990, Lew *et al* 1992, Kawka *et al* 1992, Weegels *et al* 1995, Buonocore *et al* 1996,). Wieser *et al* (1994) gives an excellent overview of the use of RP-HPLC for preparative and analytical characterization of gluten subunits.

Prediction of Quality

One of the most important reasons to analyze gluten proteins is to determine quality, i.e., functional properties that determine suitability for particular applications. Wheat quality often relates to protein amount or type; thus, gluten protein HPLC has rapidly achieved wide use for quality prediction.

RP-HPLC is an invaluable tool to estimate wheat quality. As noted above, this may be done by identifying genotypes (Bietz *et al* 1984). However, RP-HPLC can also detect specific proteins that serve as markers of quality. Bietz *et al* (1984b) first identified durum γ-gliadin RP-HPLC peaks that correlated with pasta quality and showed their correspondence to electrophoresis bands (Burnouf and Bietz 1984b). Early-generation samples could be rapidly screened for quality based on these RP-HPLC peaks. Lookhart and Albers (1988) used RP-HPLC to separate and collect each gliadin protein from sister lines varying widely in quality and then related each peak to their A-PAGE (Acid -polyacrylamide gel electrophoresis) and SDS-PAGE electrophoretic migrations and apparent size. Pattern variations were found that were attributed to possible quality differences. Many studies have since related the presence or amount of specific gliadins, individually or in combination, to quality (Lookhart and Pomeranz 1985, Huebner and Bietz 1986, Lookhart and Albers 1988, Huebner and Bietz 1988, Huebner 1989, Bunker *et al* 1989, Endo *et al* 1990, Scanlon *et al* 1990, Seilmeier *et al* 1990, Primard *et al* 1991, Huebner and Gaines 1992, Andrews *et al* 1994, Bietz and Waga 1996). Gliadin compositions vary quantitatively during kernel development and with kernel size, and may thus affect quality (Seilmeier *et al* 1990, Huebner and Gaines 1992). Quantitative differences in gliadins may also reflect differences in environment or fertilization during growth (Lookhart and Pomeranz 1985, Huebner and Bietz 1988).

Similarly, RP-HPLC has been used to relate glutenin subunits to quality. As noted above, high MW subunits related to breadmaking quality (Payne *et al* 1984) resolve well by RP-HPLC (Burnouf and Bietz 1984a, 1985); similarly, the ratio of high- to low-MW glutenin subunits relates to quality (Huebner and Bietz 1985). Additional studies have since used RP-HPLC to relate the glutenin subunit composition to quality (Marchylo *et al* 1989, Sutton *et al* 1989, 1990, Kruger and Marchylo 1990, Seilmeier *et al* 1990 Scanlon *et al* 1990, Huebner *et al* 1990, Lafiandra *et al* 1993, Hay 1993, Andrews *et al* 1994, Popineau *et al* 1994, Jia *et al* 1996). Both quantitative variation and allelic forms of high-MW subunits may influence wheat quality (Huebner *et al* 1990). RP-HPLC was used to show that dough extensibility and bread and pastry water absorption relate closely to the amount of high MW glutenin subunits (Hay 1993) and that quality deteriorates during sprouting due to degradation of high-MW subunits (Kruger and Marchylo 1990). RP-HPLC showed how quantitative variation in glutenin during wheat maturation may affect glutenin aggregation and quality (Seilmeier *et al* 1990, Jia *et al* 1996), and that loaf volumes and bake scores can be predicted by quantifying RP-HPLC results (Sutton *et al* 1989). The demonstration by Lookhart *et al* (1993b,c), Tilley *et al* (1993) and Margiotta *et al* (1996) using RP-HPLC that high-MW glutenin subunits from different varieties with the same numerical designations may not be identical is especially significant. Amounts of alcohol-soluble low-MW glutenin subunits have also been related to quality by RP-HPLC (Scanlon *et al* 1990, Andrews *et al* 1994). Computer analysis of RP-HPLC data is especially useful to identify proteins related to quality (Huebner and Bietz 1987, Bietz and Huebner 1987, Lookhart *et al* 1993a). A detailed review of RP-HPLC prediction of breadmaking quality was presented by Huebner and Bietz (1994).

Studies of Interactions and Processing

One of the most important applications of HPLC for cereal proteins may ultimately be analysis of their interactions with each other and with other kernel constituents of stored grains or grain products. As noted above, improved resolution of wheat prolamins at increased temperature suggests that RP-HPLC can monitor disruption of hydrogen bond-mediated aggregation (Bietz and Cobb 1985). Other studies suggest that non-optimal grain drying or storage is revealed by differing RP-HPLC resolution at varying temperatures. RP-HPLC may thus help optimize processing and storage conditions and maximize quality.

A few analyses of this type have been reported. Meier *et al* (1985) used RP-HPLC to analyze gliadin from untreated and heat-treated flours. For flours heated above 80-90°C, one peak was larger, indicating interaction or modification due to temperature. RP-HPLC has also been used to characterize wheat proteins that interact with lipids (Zawistowska *et al* 1986); such aggregates may form upon mixing, and their amounts correlate with bread loaf volume (Békés *et al* 1992). RP-HPLC has shown conformational changes upon disulfide cleavage or upon alkylation of cysteine residues (Bietz 1986, Burnouf and Bietz 1984a). Windemann *et al* (1986) showed that RP-HPLC can detect gliadin in heated foods, permitting monitoring of wheat and rye in foods for individuals with celiac disease. RP-HPLC of wheat proteins during mixing, fermentation, and baking has also shown degradation or interaction of specific gliadins during baking; these modifications may relate to breadmaking potential (Menkovska *et al* 1987, 1988, Pomeranz *et al* 1989). Similarly, RP-HPLC can detect high-temperature drying of pasta by changes in retention times and areas of gliadin peaks (Aktan and Khan 1992). These examples emphasize the value of HPLC for monitoring changes in protein structures or interactions, related to processing or environment, that affect end-use properties and quality.

Ion Exchange

Principles

Before HPLC, ion exchange chromatography (IEC) on cellulose columns provided the best chromatographic separation of cereal proteins. Surprisingly, however, IE-HPLC of cereal proteins has lagged behind RP- and SE-HPLC, possibly because early IEC separations were partially mixed-mode, depending upon factors besides ionic binding between solutes and support. Problems with IE-HPLC such as reproducibility and resolution have been reported, and even columns of the same type can vary among, or even within, manufacturers (Marchylo 1994). One way to reduce these problems is to purchase multiple columns of the same lot.

75

Reagents

The purity of reagents used to prepare IE-HPLC solvents is of utmost importance and the same considerations discussed above for RP-HPLC apply.

Procedure

Many solvent systems are useful for IE-HPLC of cereal proteins. Factors to consider are buffer type and pH (to convert proteins to a charged state in which they bind to ionized support surface groups), ionic strength (for the elution gradient), and use of detergents, denaturants, or organic solvents (to maintain protein solubility). Batey (1984) tested solvents at pH 8.0, 9.0, 9.5, and 10.4 for anion-exchange HPLC of gliadins. The pH 10.4 solvent (0.01 M 3-[cyclohexylamino]propanesulfonic acid) was best, apparently because many cereal proteins are deficient in negatively charged amino acids and at pH 10.4 arginine loses its positive charge. Since early ion-exchange separations of cereal proteins were generally best on cation-exchange columns, recent studies have tried cation-exchange HPLC separations. Larre et al (1991) separated gliadins with a NaCl gradient in a pH 3.3-3.8 acetate buffer containing 4 M urea. Huebner (1997) also eluted gliadins with a NaCl gradient, using pH 2.5-2.65 solvents containing 30-35% ACN and 0.045-0.5% TFA. Batey (1994) reviewed practical considerations, including buffer composition, involved in separating cereal proteins by IE-HPLC.

Standard Procedure

The following ion exchange procedure is recommended for gliadins.

1. Collumn
   Use cation exchange column, the PROT-SCX column (0.75 × 5 cm) by Vydac is particularly good.
2. Sample preparation
   Extract 100 mg sample (flour or finely ground grain) with 1 ml 70 % ethanol (or 50% propan-1-ol) in a 2 ml micro centrifuge tube by continual vortexing for 5 min and then centrifuging for 2 min at 14,000 × G. (Smaller samples can be used. As little as 10 mg can be extracted with 100 µL of solvent can be used - ½ kernel or less.) Transfer the liquid carefully into an HPLC autosampler vial. If there are problems in the separation due to the alcohol extraction solvent, the extracts can be dried by vacuum and redissolved in the HPLC starting buffer.
3. Loading
   Inject 5-10 µL (5-100 µL) into the HPLC system. Use larger volumes if you need to work with fractions collected from the separation column.

4.   Pumping conditions
    Elute at 44°C and 1 ml/min with a gradient from 0% NaCl (A), to 0.25
    M salt (B) over 50 min. Switch back to solvent A and equilibrate for
    10 min. (Longer gradients (time, %B) can be used.)
    Solvent A: 0.5 % (v/v) TFA in water buffered at pH 2.5, containing
    30 % acetonitrile.Solvent B: 0.5 % (v/v) TFA in water buffered at pH
    2.5 containing 0.25 M NaCl and 30% acetonitrile.
5.   Detection
    Record absorbance at 200, 210, 220, 254, and/or at 280 nm, (210 nm is
    preferred).

Similar gradients and conditions can be used for glutenin extracts.
Glutenins are normally extracted with 70% ethanol (or 50% propan-1-ol)
containing 1% DTT (dithiothreitol).

Varietal Identification
    Wheat varietal identification is possible by IE-HPLC (Batey 1984).
Batey (1984) first used IE-HPLC, on a Pharmacia Mono-Q anion-exchange
column, to fractionate wheat gliadins for varietal identification. More
recently, Larre *et al* (1991) fractionated gliadins by rapid cation exchange
chromatography on a Mono-S FPLC column; separation speed and
resolution were better than on a soft cation exchange gel, and genotypes
were distinguishable from their chromatographic patterns. Most recently,
Huebner achieved better separation of gliadins on a Vydac PROT-SCX
column (Fig.   7) (Huebner 1997). Undoubtedly these improvements in
cereal protein IE-HPLC, combined with the complementary nature of this
method to RP-HPLC, will lead to its increased use. Melas and Autran
(1996) fractionated glutenin subunits by IE-HPLC, and related charge
distributions to dough extensibility and baking strength.

Tips and Modifications
    Marchylo (1994) presented an excellent discussion of how to recognize
and deal with problems during gluten protein HPLC.   Some of these
important considerations and practical advice that may lead to and maintain
good separations of gluten proteins by HPLC will be discussed.
    HPLC columns for gluten protein analysis should be used according to
manufacturers' recommendations. Extreme pH must be avoided for silica,
and columns should be stored in recommended solvents. In-line filters
and/or guard columns should be used, and solvents and samples must be
freshly filtered and/or centrifuged to remove particulates. The best test of
column performance is periodic analysis of a complex, freshly-prepared
"standard" gluten protein mixture similar to samples being analyzed, such
as a 70% ethanol extract of a wheat flour.

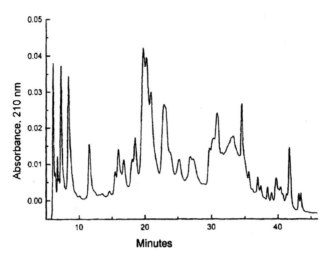

Fig. 7. IE-HPLC of gliadins from the wheat cv. Butte on a Vydac PROT-SCX column (0.75 ×5 cm). Proteins were extracted with 70% ethanol, and eluted at 44°C with a multi-step linear gradient of 0 to 0.25 M NaCl in a pH 2.5-2.65 solvent containing 30-35% ACN and 0.045-0.5% TFA. (Taken from Bietz 2002, with permission. ©Marcel Dekker, 2002).

If resolution or selectivity of the column changes, good performance may sometimes be restored by washing the column (according to manufacturers' recommendations) with solvents such as methanol, ACN, or 90% DMSO (Burnouf and Bietz 1984a). As a last resort, columns can be washed in the reverse direction; end frits can be cleaned or replaced; and packing near the column inlet can be replaced or voids filled with similar packing material. Columns do not last forever, not even newer types (Marchylo et al 1992a). Spare columns (preferably of the same lot) should be kept available.

Because of the unusual properties of many gluten proteins, sample-associated problems may arise during HPLC. Proteins may precipitate during gradient elution or at their pI as they equilibrate with a different solvent. This may be detected by monitoring resolution and pressure, and is usually prevented by guard columns. Protein solubility may also change with time due to aggregation. Samples are generally best analyzed soon after preparation; sample stability should be checked by repeat analyses over a period at least equal to that in which samples may reside in an autosampler before analysis. "Layering" phenomena may also occur as samples await analysis, making lower aliquots from a vial more concentrated. If this occurs, samples must be mixed shortly before analysis.

If precipitates form during storage of samples, they should be filtered again, or centrifuged.

A more subtle problem is presented by proteins that associate or polymerize, non-covalently or covalently, but remain soluble. Huebner and Bietz (1993) showed that medium-MW oligomeric wheat glutenin molecules eluted upon RP-HPLC as a broad peak, but resolved as sharp peaks after disulfide bond cleavage.

Most cereal proteins are rich in glutamine and are subject to deamidation in acidic solvents. It is thus best to minimize exposure to acidic solvents, including ones containing TFA (Huebner and Bietz 1987). Similarly, alkaline extraction should be avoided because of potential peptide bond hydrolysis. As noted above, high temperature may also dissociate polypeptides. Most extracts are, however, quite stable (Bietz *et al* 1984a, Burnouf and Bietz 1984a, Huebner and Bietz 1987).

Marchylo and Kruger (1988) emphasized another important consideration for RP-HPLC: injection volume. If large samples are applied in hydrophobic solvents, some less hydrophobic proteins may not bind and will elute at the void volume. This may be avoided by using solvents of low hydrophobicity, and by use of small (e.g., 5 µL) injections (multiple if necessary).

Many cereal endosperm storage proteins (especially glutelins) are, in their native forms, disulfide-bonded oligomers or polymers. Even if soluble, such proteins may resolve poorly upon RP-HPLC and elute as broad peaks (Huebner and Bietz 1993), in part due to the presence of many polymeric forms. Reduction is necessary to release disulfide-bonded subunits from polymeric or oligomeric prolamins for analysis (Burnouf and Bietz 1984a, Wieser *et al* 1989a, Kruger and Marchylo 1990). Thus, protein resolution, extractability, and stability often improve by extraction under reducing conditions. To prevent reoxidation and further enhance resolution, resulting cysteine residues may be stabilized by alkylation, usually with 4-vinylpyridine (Burnouf and Bietz 1984a, Huebner and Bietz 1987, Burnouf and Bietz 1989, Kruger and Marchylo 1990, Seilmeier *et al* 1991b). See also the chapter on Polymeric protein analyses, in this book.

It is often difficult to denature proteins completely before analysis and to assure their stability during storage, but this can be achieved by extraction with solvents containing SDS (Dachkevitch and Autran 1989, Autran 1994, Weegels *et al* 1994). Sometimes, extractants containing both denaturants (e.g., urea) and detergents (e.g., SDS) are used to dissociate both hydrogen and hydrophobic bonds. Sonication is also effective to solubilize native polymeric glutelins (Singh *et al* 1990).

The need for reproducible HPLC separations within and between laboratories is complicated by different selectivities or resolution of columns varying in bonded phase, source, and age. To correct for such differences and monitor system performance, standardization is necessary. The problem in RP-HPLC is difficult. Ideally, standardization between laboratories or between columns would use stable, commercially-available compounds as standards. Bietz and Cobb (1985) tested alkylphenones as RP-HPLC standards, but use of such compounds has never been widely practiced, and few other suitable standards are available. More typically, within a laboratory, the best standard may be a fresh, easily prepared heterogeneous sample, such as a 70% ethanol extract of wheat flour, which gives a characteristic and complex chromatogram. Changes in resolution, retention times, or selectivity that indicate problems with the column or system are easily seen. Such procedures have been used to ensure long-term reproducibility of analyses, to normalize data based on characteristic peaks, to compare samples, and to automatically identify genotypes from their characteristic fingerprints (Sapirstein et al 1989, Scanlon et al 1989a,b, 1990).

To compare samples between laboratories and develop standard methods, agreement as to optimal extraction and analysis methods is desirable. The diverse aims and nonstandard nature of most separations, combined with inevitable changes in columns and optimization of procedures, make such a goal elusive. A collaborative study to determine optimal RP-HPLC analytical conditions for wheat proteins (Lookhart and Bietz 1994) suggested that adoption of standard methods is unlikely, but showed that the optimal conditions for wheat proteins are similar to those for other proteins. Procedures and results from different laboratories were also similar. For RP-HPLC, ACN-containing solvents (gradients from 15 to 70% ACN, + 0.05-0.1% TFA) are most common, with detection at ca. 210 nm to maximize sensitivity. Fixed-wavelength UV detection is normally used for proteins; photodiode-array detection at multiple wavelengths does not improve wheat varietal identification since most gliadins have similar spectral characteristics (Scanlon and Bushuk 1990). C_4, C_8, and C_{18} phases bonded to ca. 5 μm spherical porous (ca. 300 A) silica particles are most effective, and sterically-protected silica-based columns offer improved stability (Marchylo et al 1992a). Columns of 4-5 mm i.d., 15-25 cm long, are most common, but shorter, narrow-bore columns are increasingly used (Huebner and Bietz 1995). Separation times of 50 to 120 minutes generally give optimal resolution with 4-5 mm i.d. columns, but narrow- or micro-bore columns often give comparable separation efficiency in 10 to 20 minutes or less (Huebner and Bietz 1995). Most investigators use temperatures of 60-70°C for RP-HPLC. Using these conditions, RP-HPLC offers high resolution, sensitivity, and reproducibility; procedures are easy to learn and automate, and results can be readily quantified.

Conclusions

HPLC methods have, during the last 15 years, become widely used and are now methods of choice for many cereal protein analyses. RP-HPLC is especially useful because of its speed, sensitivity, high resolution, reproducibility, and quantifiability, and because it complements other methods. IE-HPLC has developed more slowly, but is becoming useful. This chapter has reviewed these methods with special emphasis on their use to characterize cereal proteins; to identify varieties, to act as preparative tools, to determine varietal purity, to analyze hybrids and mixtures, to be used as selection tools during breeding; to function as quantitative genetic tools, to predict quality, and to study interactions and conformations. HPLC methods are now essential tools for understanding, improving, and exploiting the full potential of one of our most important agricultural commodities and the products derived from it.

References

Aguilar, M.I. and Hearn, M.T.W. 1996. High-resolution reversed-phase high-performance liquid chromatography of peptides and proteins. Pp 3-26 in Meth. Enzymol., 270. (B.L. Karger and W. S. Hancock, eds.), Academic Press, San Diego, CA.

Aibara, S. and Morita, Y. 1988. Purification and characterisation of wheat gamma-gliadins. J. Cereal Sci. 7:237.

Aktan, B. and Khan, K. 1992. It's effect of high-temperature drying of pasta on quality parameters and on solubility, gel electrophoresis, and reversed-phase high-performance liquid chromatography of protein components. Cereal Chem. 69:288.

Andrews, J.L., Hay, R.L., Skerritt, J.H. and Sutton, K.H. 1994. Effect of high-temperature drying of pasta on quality parameters and on solubility, gel electrophoresis, and reversed-phase high-performance liquid chromatography of protein components. J. Cereal Sci. 20:203.

Autran, J.-C. 1994. Size-exclusion high-performance liquid chromatography for rapid examination of size differences of cereal proteins, p.326 in High-Performance Liquid Chromatography of Cereal and Legume Proteins (J.E. Kruger and J.A. Bietz, eds.), Am. Assoc. Cereal Chem. St. Paul, MN.

Barnwell, P., Mccarthy, P.K., Lumley, I.D. and Griffin, M. 1994. The use of reversed-phase high-performance liquid chromatography to detect common wheat (Triticum aestivum) adulteration of durum wheat (Triticum durum) pasta products dried at low and high temperatures. J. Cereal Sci. 20:245.

Batey, I. 1984. Wheat varietal identification by rapid ion-exchange chromatography of gliadins. J. Cereal Sci. 2:241.

Batey, I.L. 1994. HPLC Ion-Exchange Chromatographic Separations of Cereal and Legume Proteins, p.373 in High-Performance Liquid Chromatography of Cereal and Legume Proteins (J.E. Kruger and J.A. Bietz, eds.), Am. Assoc. Cereal Chem. St. Paul, MN.

Bean, S.R. and Lookhart, G.L. 1997. Separation of wheat proteins by two-dimensional reversed-phase high-performance liquid chromatography plus free zone capillary electrophoresis. Cereal Chem. 74:758.

Békés, F., MacRitchie, F., Panozzo, J.F., Batey , I.L. and O'Brien, L. 1992. Lipid mediated aggregates in flour and in gluten. J. Cereal Sci. 16:129.

Bietz, J.A. 1983. Separation of cereal proteins by reversed-phase high-performance liquid chromatography. J. Chromatogr. 255:219.

Bietz, J.A. 1986. High-performance liquid chromatography of cereal proteins. p. 105 in Advances in Cereal Science and Technology, Vol. 8 (Y. Pomeranz, ed.), Am. Assoc. Cereal Chem. St. Paul, MN.

Bietz, J.A. 2002. HPLC of cereal endosperm storage proteins. Pages 547-587 in: HPLC of Biological Macromolecules, 2nd ed., revised and expanded. K M. Gooding and F. E. Regnier, eds. Marcel Dekker, Inc., New York.

Bietz, J.A. and Cobb, L.A. 1985. Improved procedures for rapid wheat varietal identification by reversed-phase high-performance liquid chromatography of gliadin. Cereal Chem. 62:332.

Bietz, J.A. and Huebner, F.R. 1987. Prediction of wheat quality by computer evaluation of reversed-phase high-performance liquid chromatograms of gluten proteins, p.173 in Proceedings, 3rd International Workshop on Gluten Proteins (R. Lásztity and F. Békés, eds.), Budapest Hungary, World Scientific, Singapore.

Bietz, J.A. and Huebner, F.R. 1994. Variety identification by HPLC, p.73 in Identification of Food-Grain Varieties (C. W. Wrigley, ed.), Am. Assoc. Cereal Chem. St. Paul, MN.

Bietz, J.A. and Waga, J. 1996. Gliadin block analysis by RP-HPLC and capillary electrophoresis for genotype identification and quality prediction, p.362 in Gluten '96 - Proc. 6th Int. Gluten Workshop (C. W. Wrigley, ed.), Cereal Chem. Div., Royal Australian Chemical Inst., North Melbourne, Australia.

Bietz, J.A., Burnouf, T., Cobb, L.A. and Wall, J.S. 1983. Differences in selectivity of large-pore reversed-phase high-performance liquid chromatography columns for analysis of cereal proteins. Cereal Foods World, 28:555.

Bietz, J.A., Burnouf, T., Cobb, L.A., and Wall, J.S. 1984a. Gliadin analysis by reversed-phase high-performance liquid chromatography: Optimization of extraction conditions. Cereal Chem. 61:124.

Bietz, J.A., Burnouf, T., Cobb, L.A. and Wall, J.S. 1984b. Wheat varietal identification and genetic analysis by reversed-phase high-performance liquid chromatography. Cereal Chem. 61:129.

Bunker, J.R., Lockerman, R.H,. Mcguire, C.F., Blake, T.K. and Engel, R.E. 1989. Soil Moisture effects on bread loaf quality and evaluation of gliadins with reversed-phase high-performance liquid chromatography. Cereal Chem. 66:427.

Buonocore, F., Caporale, C. and Lafiandra, D. 1996. Purification of characterization of high M_r glutenin subunit 20 and its linked y-type subunit from durum wheat. J. Cereal Sci. 23:195.

Burnouf, T. and Bietz, J.A. 1984a. Reversed-phase high-performance liquid chromatography of reduced glutenin, a disulfide-bonded protein of wheat endosperm. J. Chromatogr. 299:185.

Burnouf, T. and Bietz, J.A. 1984b. Reversed-phase high-performance liquid chromatography of durum wheat gliadins: Relationships to durum wheat quality. J. Cereal Sci. 2:3.

Burnouf, T. and Bietz, J.A. 1985. Chromosomal control of glutenin subunits in aneuploid lines of wheat: Analysis by reversed-phase high-performance liquid chromatography. Theor. Appl. Genet. 70: 610.

Burnouf, T. and Bietz, J.A. 1989. Rapid purification of wheat glutenin for reversed-phase high-performance liquid chromatography: Comparison of dimethyl sulfoxide with traditional solvents. Cereal Chem. 66:121.

Burnouf, T., Bietz, J.A., Cobb, L.A. and Wall, J.S. 1983. Reversed-phase high-performance liquid chromatography of gliadins from hexaploid wheat (Triticum aestivum) varieties cultivated in France and possibility of varietal identification.. C. R. Acad. Sci. Paris, 297:377.

Chen, H. and Horvath, C. 1995. High-speed high-performance liquid chromatography of peptides and proteins. J. Chromatogr. 705:3.

Chloupek, R.C., Hancock, W.S., Marchylo, B.A., Kirkland, J.J., Boyes, B.E. and Snyder, 1994. Temperature as a variable in reversed-phase high-performance liquid chromatographic separations of peptide and protein samples. II. Selectivity effects observed in the separation of several peptide and protein mixtures. J. Chromatogr. 686:45.

Cressey, P.J. 1987. Identification of New Zealand wheat cultivars by reversed-phase high-performance liquid chromatography. N. Z. J. Agric. Res. 30:125.

Dachkevitch, T. and Autran, J.-C. 1989. Prediction of baking quality of bread wheats in breeding programs by size-exclusion high-performance liquid chromatography. Cereal Chem. 66:448.

De Noni, I., De Bernardi, G. and Pellegrino, L. 1994. Detection of common-wheat (Triticum aestivum) flour in durum-wheat (Triticum durum) semolina by reversed-phase high-performance liquid chromatography.(Rp-hplc) of specific albumins. Food Chem. 51:325.

Endo, S., Okada, K., Nagao, S. and D'Appolonia, B.L. 1990. Quality characteristics of hard red spring and winter wheats. I. Differentiation by reversed-phase high-performance liquid chromatography and milling properties. Cereal Chem. 67:480.

Freedman, A.R., Wieser, H., Ellis, H.J. and Ciclitira, P.J. 1988. Immunoblotting of gliadins separated by reversed-phase high-performance liquid chromatography: Detection with monoclonal antibodies. J. Cereal Sci. 8:231.

Gupta, R.B., Masci, S., Lafiandra, D., Bariana, H.S., and MacRitchie, F. 1996. Accumulation of protein subunits and their polymers in developing grains of hexaploid wheat. J. Exp. Bot. 47:1377.

Hansen, V. and Prentice, 1985. Reversed-phase high-performance liquid chromatography analytical method for winter wheat variety identification. Newsl. Assoc. Off. Seed Analysts, 59:103.

Hay, R.L. 1993. Effect of flour quality characteristics on puff pastry baking performance. Cereal Chem. 70:392.

Hay, R.L. and Sutton, K.H. 1990. Identification and discrimination of New Zealand bread wheat, durum wheat, and rye cultivars by RP-HPLC. New Zealand J. Crop and Hortic. Sci. 18:49.

Hearn, M.T.W. 1980. The use of reversed-phase high-performance liquid chromatography for the structural mapping of polypeptides and proteins. J. Liq. Chromatogr. 3:1255.

Huebner, F.R. 1989. Assessment of potential breadmaking quality of hard spring wheats by high-performance liquid chromatography of gliadins - Year two. Cereal Chem. 66:333.

Huebner, F.R. 1997. Ion-exchange HPLC of cereal proteins. Cereal Foods World. 42:613.

Huebner, F.R. and Bietz, J.A. 1984. Separation of wheat gliadins by preparative reversed-phase high-performance liquid chromatography. Cereal Chem. 61:544.

Huebner, F.R. and Bietz, J.A. 1985. Detection of quality differences among wheats by high-performance liquid chromatography. J. Chromatogr. 327:333.

Huebner, F.R. and Bietz, J.A. 1986. Assessment of the potential breadmaking quality of hard wheats by reversed-phase high-performance liquid chromatography of gliadins. J. Cereal Sci. 4:379.

Huebner, F.R. and Bietz, J.A. 1987. Improvements in wheat protein analysis and quality prediction by reversed-phase high-performance liquid chromatography. Cereal Chem. 64:15.

Huebner, F.R. and Bietz, 1988. Quantitative variation among gliadins of wheats grown in different environments. Cereal Chem. 65:362.

Huebner, F.R. and Bietz, J.A. 1993. Improved chromatographic separation and characterization of ethanol-soluble wheat proteins. Cereal Chem. 70:506.

Huebner, F.R. and Bietz, J.A. 1994. RP-HPLC for assessment of quality in cereals and legumes. Breadmaking quality (wheat), p.206 in High-Performance Liquid Chromatography of Cereal and Legume Proteins (J.E. Kruger and J.A. Bietz, eds.), Am. Assoc. Cereal Chem. St. Paul, MN.

Huebner, F.R. and Bietz, J.A. 1995. Rapid and sensitive wheat protein fractionation and varietal identification by narrow-bore reversed-phase high-performance liquid chromatography. Cereal Chem. 72:504.

Huebner, F.R. and Gaines, C.S. 1992. Relationship between wheat kernel hardness, environment, and gliadin composition. Cereal Chem. 69:148.

Huebner, F.R., Kaczkowski, J., and Bietz, J.A. 1990. Quantitative variation of wheat proteins from grain at different stages of maturity and from different spike locations. Cereal Chem. 67:464.

Huebner, F.R. , Nelsen, T. C. and Bietz, J.A. 1995. Differences among gliadins from Spring and Winter wheat cultivars. Cereal Chem. 72:341.

Jia, Y.Q., Masbou, V., Aussenac, T., Fabre, J.L. and Debaeke, P. 1996. Effects of nitrogen fertilization and maturation on protein aggregates and on the breadmaking quality of Soissons, a common wheat cultivar. Cereal Chem. 73:123.

Jones, B.L., Lookhart, G.L. and Johnson, D.E. 1985. Improved separation and toxicity analysis methods for purothionins. Cereal Chem. 62: 327.

Kawka, A., Ng, P.K.W. and Bushuk, W. 1992. Equivalence of high molecular weight glutenin subunits prepared by reversed-phase high-performance liquid chromatography and sodium dodecyl sulfate-polyacrylamide gel electrophoresis. Cereal Chem. 69:92 .

Khan, K., Huckle, L. and Jones, B.L. 1992. Inheritance of gluten components of a high-protein hard red spring wheat line derived from Triticum turgidum Var. Dicoccoides - Semipreparative RP-HPLC, gel electrophoresis, and amino acid composition studies. Cereal Chem. 69:270.

Kruger, J.E. and Marchylo, B.A. 1984. Selection of column and operating conditions for reversed-phase high-performance liquid chromatography of proteins in Canadian wheat. Can. J. Plant Sci. 65:285.

Kruger, J.E. and Marchylo, B.A. 1985. Examination of the mobilization of storage proteins of wheat kernels during germination by high-performance reversed-phase and gel permeation chromatography. Cereal Chem. 62:1.

Kruger, J.E. and Marchylo, B.A. 1990. Analysis by reversed-phase high-performance liquid chromatography of changes in high molecular weight subunit composition of wheat storage proteins during germination. Cereal Chem. 67:141.

Kruger, J.E., Marchylo, B.A. and Hatcher, D. 1988. Preliminary assessment of a sequential extraction scheme for evaluating quality by reversed-phase high-performance liquid chromatography and electrophoretic analysis of gliadins and glutenins. Cereal Chem. 65:208.

Lafiandra, D, D'Ovidio, R., Porceddu, E., Margiotta, B. and Colaprico, G. 1993. New data supporting high Mr glutenin subunit 5 as the determinant of quality differences among the pairs 5 + 10 vs 2 + 12. J. Cereal Sci. 18:197.

Larre, C., Popineau, Y. and Loisel, W. 1991. Fractionation of gliadins from common wheat by cation exchange FPLC. J. Cereal Sci. 14:231.

Lew, E. J.-L., Kuzmicky, D.D. and Kasarda, D.D. 1992. Characterization of low molecular weight glutenin subunits by reversed-phase high-performance liquid chromatography, sodium dodecyl sulfate-polyacrylamide gel electrophoresis, and N-terminal amino acid sequencing. Cereal Chem. 69:508.

Lookhart, G.L. 1997. New methods helping to solve the gluten puzzle. Cereal Foods World 42:16.

Lookhart, G.L. and Albers, L.D. 1988. Correlations between reversed-phase high-performance liquid chromatography and acid- and sodium dodecyl sulfate-polyacrylamide gel electrophoretic data on prolamins from wheat sister lines differing widely in baking quality. Cereal Chem. 65:222.

Lookhart, G.L., Albers, L.D., and Bietz, J.A. 1986. A comparison of PAGE and HPLC methods for analysis of gliadin polymorphism in the wheat cultivar `Newton'. Cereal Chem. 63:497.

Lookhart, G.L. and Bietz, J.A. 1990. Practical wheat variety identification in the United States. Cereal Foods World 35:404.

Lookhart, G.L. and Bietz, J.A. 1994. Protein extraction and sample handling techniques, p.51 in High-Performance Liquid Chromatography of Cereal and Legume Proteins (J.E. Kruger and J.A. Bietz, eds.), Am. Assoc. Cereal Chem. St. Paul, MN.

Lookhart, G.L., Cox, T.S., and Chung, O.K. 1993a. Statistical analyses of gliadin reversed-phase high-performance liquid chromatography (RP-HPLC) patterns of hard red spring and hard red winter wheat cultivars grown in a common environment: Classification indices. Cereal Chem. 70:430.

Lookhart, G.L., Graybosch, R., Peterson, J., and Lukaszewski, A. 1991. Identification of wheat lines containing the 1BL/1RS translocation by high performance liquid chromatography (HPLC). Cereal Chem. 68:312.

Lookhart, G.L., Hagman, K., and Kasarda, D.D. 1993b. High-molecular-weight glutenin subunits of the most commonly grown wheat cultivars in the U.S. in 1984. Plant Breeding 110:48-62.

Lookhart, G.L., Lai, F.S., and Pomeranz, Y. 1985. Variability in gliadin electrophoregrams and hardness of individual wheat kernels selected from foundation seed on the basis of grain morphology. Cereal Chem. 62:185.

Lookhart, G.L., Marchylo, B.A., Khan, K., Lowe, D.B., Mellish, V.J., and Seguin, L. 1994. Wheat identification in North America, pp. 201-208 in Wrigley, C.W. (ed.) Identification of Food Grain Varieties, Am. Assoc. Cereal Chem. St. Paul, MN.

Lookhart, G.L., Martin, M. L., Mosleth, E., Uhlen, A.K., and Hoseney, R.C. 1993c. Comparison of high-molecular-weight subunits of glutenin and baking performance of flours varying in bread-making quality. Food Sci. Tech. 26:301-306.

Lookhart, G.L., Menkovska, M., and Pomeranz, Y. 1989. Polyacrylamide gel electrophoresis and high-performance liquid chromatography patterns of gliadins from wheat sections and milled and air-classified fractions. Cereal Chem. 66:256.

Lookhart, G.L. and Pomeranz, Y. 1985. Gliadin high-performance liquid chromatography and polyacrylamide gel electrophoresis patterns of wheats grown with fertilizer treatments in the United States and Australia on sulfur-deficient soils. Cereal Chem. 62:227.

Mahoney, W.C. and Hermodson, M.A. 1980. Separation of large denatured peptides by reversed-phase high-performance liquid chromatography. Trifluoroacetic acid as a peptide solvent. J. Biol. Chem. 255: 11199.

Marchylo, B.A. 1994. Practical considerations in the RP-HPLC analysis of cereal and legume proteins, p.15 in High-Performance Liquid Chromatography of

Cereal and Legume Proteins (J.E. Kruger and J.A. Bietz, eds.), Am. Assoc. Cereal Chem. St. Paul, MN.

Marchylo, B.A., Hatcher, D.W. and Kruger, J.E. 1988. Identification of wheat cultivars by reversed-phase high-performance liquid chromatography of storage proteins. Cereal Chem. 65:28.

Marchylo, B.A., Hatcher, D.W., Kruger, J.E. and Kirkland, J.J. 1992a. Reversed-phase high-performance liquid chromatographic analysis of wheat proteins using a new, highly stable column. Cereal Chem. 69:371.

Marchylo, B.A. and Kruger, J.E. 1988. The effect of injection volume on the quantitative analysis of wheat storage proteins by reversed-phase high-performance liquid chromatography. Cereal Chem. 65:192.

Marchylo, B.A., Kruger, J.E. and Hatcher, D.W. 1989. Quantitative Reversed-phase high-performance liquid chromatographic analysis of wheat storage proteins as a potential quality prediction tool. J. Cereal Sci. 9:113 .

Marchylo, B.A., Kruger, J.E. and Hatcher, D.W. 1990. Effect of environment on wheat storage proteins as determined by quantitative reversed-phase high-performance liquid chromatography. Cereal Chem. 67:372.

Marchylo, B.A., Lukow, O.M. and Kruger, J.E. 1992b. Quantitative variation in high molecular weight glutenin subunit 7 in some Canadian wheat. J. Cereal Sci. 15:29.

Margiotta, B., Colaprico, G., D'Ovidio, R. and Lafiandra, D. 1993. Characterization of high Mr subunits of glutenin by combined chromatographic (RP-HPLC) and electrophoretic separation and restriction fragment length polymorphism (RFLP) analyses of their encoding genes. J. Cereal Sci. 17:221.

Margiotta, B., Urbano, M. ,Colaprico, ,G., Johansson, E., Buonocore, F., D'Ovidio, R. and Lafiandra, D. 1996. Detection of y-type subunit at the Glu-A1 locus in some Swedish bread wheat lines. J. Cereal Sci. 23:203.

Masci, S., Lew, E. J.-L., Lafiandra, D., Porceddu, E. and Kasarda, D.D. 1995. Characterization of low molecular weight glutenin subunits in durum wheat by reversed-phase high-performance liquid chromatography and N-terminal sequencing. Cereal Chem. 72:100.

McCarthy, P.K., Cooke, R.J., Lumley, I.D., Scanlon, B.F. and Griffin, M. 1990a. Application of reversed-phase high-performance liquid chromatography for the estimation of purity in hybrid wheat. Seed Sci. and Technol., 18:609.

McCarthy, P.K. , Scanlon, B.F., Lumley, I.D. and Griffin, M. 1990b. Detection and quantification of adulteration of durum wheat flour by flour from common wheat using reversed-phase HPLC. J. Sci. Food Agric. 50:211.

Meier, P., Windemann, H. and Baumgartner, E. 1985. Analysis of whole gliadin from untreated and heat-treated wheat flours by reversed-phase high-performance liquid chromatography. Z. Lebensmittel-Untersuch. Fors. 180:467.

Melas, V. and Autran, J-C. 1996. Biochemical and functional characterization of soft wheat glutenins by cation exchange chromatography. Relation with the technological properties of dough. Sci. des Aliments. 16:361.

Menkovska, M., Lookhart, G.L., and Pomeranz, Y. 1987. Changes in the gliadin fractions during breadmaking: Isolation and characterization by HPLC and PAGE. Cereal Chem. 64:311.

Menkovska, M., Pomeranz, Y., Lookhart, G.L. and Shogren, M.D. 1988. Gliadin in crumb of bread from high-protein wheat flours of varied breadmaking potential. Cereal Chem. 65:198.

Muller, S. and Wieser, H. 1995. The location of disulfide bonds in alpha-type gliadins. J. Cereal Sci. 22:21.

Neville, B. 1996. Reversed-phase chromatography of proteins, p.277 in Protein Purification Protocols (S. Doonan, ed.), Humana Press Inc., Totowa, NJ.

Ng, P.K.W., Scanlon, M.G. and Bushuk, W. 1989. Electrophoretic and high-performance liquid chromatography patterns of registered Canadian what cultivars. Cereal Res. Commun. 17:5.

Ooms, B. 1996. Temperature control in high-performance liquid chromatography. LC-GC, 14:306.

Payne, P.I., Holt, L.M., Jackson, E.A. and Law, C.N. 1984. Wheat storage proteins: Their genetics and their potential for manipulation by plant breeding. Phil. Trans. Roy. Soc. London B, 304:359.

Pomeranz, Y., Lookhart, G.L., Rubenthaler, G.L. and Albers, L. 1989. Changes in gliadin proteins during cookie making. Cereal Chem. 66:532.

Popineau, Y. 1994. Evaluation of hydrophobicity of wheat proteins and peptides by HIC and RP-HPLC, p.393 in High-Performance Liquid Chromatography of Cereal and Legume Proteins (J.E. Kruger and J.A. Bietz, eds.), Am. Assoc. Cereal Chem. St. Paul, MN.

Popineau, Y., Cornec, M., Lefebvre, J. and Marchylo, B. 1994. Evaluation of hydrophobicity of wheat proteins and peptides by HIC and RP-HPLC. J. Cereal Sci. 19:231.

Popineau, Y. and Pineau, F. 1987. Investigation of surface hydrophobicities of purified gliadins by hydrophobic interaction chromatography, reversed-phase high-performance liquid chromatography, and apolar binding. J. Cereal Sci. 5:215.

Primard, S., Graybosch, R., Peterson, C.J. and Lee, J.-H. 1991. Relationships between gluten protein composition and quality characteristics in four populations of high-protein, hard red winter wheat. Cereal Chem. 68:305.

Ram, C., Huebner, F.R. and Bietz, J.A. 1995. Identification of Indian wheat varieties by reversed-phase high-performance liquid chromatography. Seed Sci. and Technol., 23:259.

Sapirstein, H.D., Scanlon, M.G. and Bushuk, W. 1989. Normalization of high-performance liquid chromatography peak retention times for computerized comparison of wheat prolamin chromatograms. J. Chromatogr. 469:127.

Scanlon, M.G. and Bushuk, W. 1990. Application of photodiode-array detection in RP-HPLC of gliadins for automated wheat variety identification. J. Cereal Sci. 12:229.

Scanlon, M.G., Ng, P.K.W., Lawless, D.E. and Bushuk, W. 1990. Suitability of reversed-phase high-performance liquid chromatographic separation of wheat proteins for long-term statistical assessment of breadmaking quality. Cereal Chem. 67:395.

Scanlon, M.G., Sapirstein, H.D. and Bushuk, W. 1989a. Computerized wheat varietal identification by high-performance liquid chromatography. Cereal Chem. 66:112.

Scanlon, M.G., Sapirstein, H.D. and Bushuk, W. 1989b. Computerized wheat varietal identification by high-performance liquid chromatography. Cereal Chem. 66:439.

Seilmeier, W., Belitz, H-D. and Wieser, H. 1991a. Separation and quantitative determination of high-molecular-weight subunits of glutenin from different wheat varieties and genetic variants of the variety Sicco. Z. Lebensmittel-Untersuch. Fors. 192:124.

Seilmeier, W., Belitz, H-D. and Wieser, H. 1991b. Studies of glutenin subunits from different wheat varieties, p.287 in Gluten Proteins 1990 (W. Bushuk and R. Tkachuk, eds.), Am. Assoc. Cereal Chem. St. Paul, MN.

Seilmeier, W., Wieser, H. and Belitz, H-D. 1987. High-performance liquid chromatography of reduced glutenin: Amino acid composition of fractions and components. Z. Lebensmittel-Untersuch. Fors. 185:487.

Seilmeier, W., Wieser, H. and Belitz, H-D. 1988. Reversed-phase high-performance liquid chromatography of reduced glutenins from different wheat varieties. Z. Lebensmittel-Untersuch. Fors. 187:107.

Seilmeier, W., Wieser, H. and Belitz, H-D. 1990. Wheat during maturation: Analysis of gliadins and glutenins by RP-HPLC. Z. Lebensmittel-Untersuch. Fors. 191:99.

Singh, N.K., Donovan, G.R., Batey, I.L. and MacRitchie, F. 1990. Use of sonication and size-exclusion high-performance liquid chromatography in the study of wheat flour proteins. I. Dissolution of total proteins in the absence of reducing agents. Cereal Chem. 67:150.

Sutton, K.H. 1991. Qualitative and quantitative variation among high-molecular weight subunits of glutenin detected by reversed-phase high-performance liquid chromatography. J. Cereal Sci. 14:25.

Sutton, K.H., and Hay, R. L. 1990. Quantitation of rye in wheat/rye wholemeal mixtures by reversed-phase high-performance liquid chromatography. J. Cereal Sci. 12:25.

Sutton, K.H., Hay, R.L. and Griffin, W.B. 1989. Assessment of the potential bread baking quality of New Zealand wheats by RP-HPLC of glutenins. J. Cereal Sci. 10:113.

Sutton, K.H., Hay, R.L., Mouat, C.H. and Griffin, W.B. 1990. The influence of environment, milling, and blending on assessment of the potential breadbaking quality of wheat by RP-HPLC of glutenin subunits. J. Cereal Sci. 12:145.

Tilley, K.A., Lookhart, G.L., Hoseney, R.C. and Mawhinney, T.P. 1993. Evidence for glycosylation of the high molecular weight glutenin subunits 2, 7, 8, and 12 from Chinese Spring and TAM 105 wheats. Cereal Chem. 70:602-606.

Vensel, W.H., Lafiandra, D. and Kasarda, D.D. 1989. The effect of an organic eluent modifier and pH on the separation of wheat-storage proteins: Application to the purification of gamma-gliadins of Triticum monococcum L. Chromatographia 28: 133.

Weegels, P.L., Flissebaalje, T., and Hamer, R.J. 1994. Factors affecting the extractability of the glutenin macropolymers. Cereal Chem. 71:308.

Weegels, P.L., Hamer, R.J. and Schofield, J.D. 1995. RP-HPLC and capillary electrophoresis of subunits from glutenin isolated by SDS and Osborne fractionation. J. Cereal Sci. 22:211.

Werbeck, U., Seilmeier, W. and Belitz, H-D. 1989. Partial reduction of wheat glutelin by mercaptoethanol and dithioerythritol. II. Investigation by reversed-phase high-performance liquid chromatography. Z. Lebensmittel-Untersuch. Fors. 188:22.

Wieser, H. and Belitz, H-D. 1992. Coeliac active peptides from gliadin: Large-scale preparation and characterization. Z. Lebensmittel-Untersuch. Fors. 194:229.

Wieser, H., Modl, A., Seilmeier, W. and Belitz, H-D. 1987. High-performance liquid chromatography of gliadins from different wheat varieties: Amino acid composition and N-terminal amino acid sequencing of components. Z. Lebensmittel-Untersuch. Fors. 185:371.

Wieser, H., Seilmeier, W. and Belitz, H-D. 1989a. Reversed-phase high-performance liquid chromatography of ethanol-soluble and ethanol-insoluble reduced glutenin fractions. Cereal Chem. 66:38.

Wieser, H.,Seilmeier, W. and Belitz, H-D. 1989b. Extractability of glutenins with water/alcohol mixtures under reducing conditions. Z. Lebensmittel-Untersuch. Fors. 189:223.

Wieser, H., Seilmeier, W. and Belitz, H-D. 1990. Fractionation, separation, and characterization of reduced glutenins from the variety Rektor. II. characterization of high-molecular-weight subunits of glutenin separated by reversed-phase high-performance liquid chromatography. J. Cereal Sci. 12:63 .

Wieser, H., Seilmeier, W. and Belitz, H-D. 1994. Use of RP-HPLC for a better understanding of the structure and functionality of wheat gluten proteins, p.235 in High-Performance Liquid Chromatography of Cereal and Legume Proteins (J.E. Kruger and J.A. Bietz, eds.), Am. Assoc. Cereal Chem. St. Paul, MN.

Windemann, H., Meier, P. and Baumgartner, E. 1986. Detection of wheat gliadins in heated foods by reversed-phase high-performance liquid chromatography. Z. Lebensm Unters. Forsch. 183:26 .

Zawistowska, U., Bietz, J.A. and Bushuk, W. 1986. Characterization of low-molecular-weight protein with high affinity for flour lipid from two wheat cultivars. Cereal Chem. 63:414.

Chapter 5

Separation of Gluten Proteins by High Performance Capillary Electrophoresis

Scott R. Bean
USDA-ARS, Grain Marketing and Production Research Center
Manhattan, Kansas

George L. Lookhart
USDA-ARS, Grain Marketing and Production Research Center
Manhattan, Kansas

Introduction

In the Greek language protein means "of prime importance" (Hegarty 1995). Gluten proteins are particularly suited for this title because of their ability to form visco-elastic doughs (MacRitchie 1992). Because of this unique ability, gluten proteins have been intensively studied by a number of analytical techniques. This chapter will focus on a relative newcomer to the field of cereal chemistry, high performance capillary electrophoresis (HPCE). HPCE utilizes small inner diameter capillaries as an anti-convective medium in place of slab-gels. Due to the small inner diameter of these capillaries (typically 50 to 100 μm) high voltages can be used, resulting in rapid, high resolution separations. More comprehensive reviews on HPCE are available (Guzman 1993, Righetti 1996, Landers 1997, Wehr *et al* 1999).

Like traditional slab-gel electrophoresis, HPCE can operate in several modes. Methods for two modes have been developed for separating wheat proteins. These two modes are free zone capillary electrophoresis (FZCE), and sodium dodecyl sulfate capillary electrophoresis (SDS-CE).

U.S. Department of Agriculture, Agricultural Research Service, Northern Plains Area, is an equal opportunity/affirmative action employer and all agency services are available without discrimination.
Mention of firm names or trade products does not constitute endorsement by the U.S. Department of Agriculture over others not mentioned.

91

FZCE

Principles

FZCE is the simplest and most often used mode of HPCE. In this mode of HPCE, capillaries are simply filled with separation buffer and proteins are separated by differences in their charge density. Most (but not all) successful separations of wheat proteins by FZCE have been carried out at acidic pH making this technique analogous to separation of wheat proteins by acid polyacrylamide gel electrophoresis (A-PAGE).

In FZCE, capillaries are flushed with one or more conditioning rinses and then rinsed and filled with separation buffer. After the capillary is properly conditioned, sample is introduced into the capillary, typically by pressure injection though other modes can also be used. The ends of the capillary are then covered by buffer vials and the voltage is turned on. This establishes an electrical field inside the capillary. Proteins then migrate to the appropriate electrode. As mentioned previously, most separations of gluten proteins by FZCE have been carried out at acidic pH where the proteins carry net positive charges. Thus during FZCE separation the proteins migrate towards the cathode.

Proteins are typically detected by UV light in FZCE separations. This is accomplished by shining UV light through a small portion of the capillary where the protective polyimide coating has been removed. This "detection window" is normally placed near one end of the capillary. Thus in order to detect the proteins, they must actually pass through the part of the capillary containing the detection window. Therefore it is critically important to set the polarity correctly on the instrument so that the proteins migrate towards the detection window. In addition to UV detection, many other detection types can be used such as mass spectrometry and laser induced fluorescence. A diagram of a wheat protein separation by FZCE at acidic pH is shown in Fig. 1.

Reagents

All reagents used for FZCE separations should be of the highest possible purity (e.g. Sigma Ultrapure). Impurities in reagents used in buffer preparation can cause increased currents, may bind to capillary walls, or possibly interact with proteins themselves.

Separation Buffer Preparation

The buffer utilized by this method consists of 50 mM IDA, 20% acetonitrile (ACN) and 0.05% hydroxypropylmethyl-cellulose (HPMC) (with a viscosity of a 2% solution = 4000 cps). To prepare 500 mL of this buffer the following procedure is used:

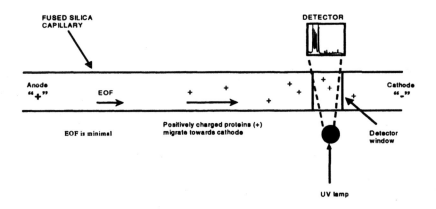

Fig. 1. Diagram of a FZCE separation of gluten proteins at acidic pH.

1. 3.38 g IDA, 100 mL UV grade ACN, and 20 mL of 1.25% HPMC solution are weighed out and mixed together with ~250 mL of high purity deionized water.
2. After the chemicals are completely dissolved the volume is adjusted to 500 mL with high purity deionized water. No further preparation is necessary.
3. For best results, small fractions should be aliquoted out (50-100 mL) and stored in a refrigerator until needed. Buffer can be safely stored for months, but the initial current should be noted when a new batch of buffer is made. Over time this current level may increase due to evaporation of the ACN and subsequent concentration of the buffer.

Extraction Solvents

Albumin and globulin solvent
    The solvent to extract albumin and globulin proteins is a 50 mM TrisHCl buffer, pH 7.8 containing 100 mM KCl and 5 mM EDTA (Bean and Lookhart 1998).
1. Dissolve 3.7 g KCl, 2.7 g TrisHCl, 1.0 g Tris, and 0.7 g EDTA in 250 mL high purity water.
2. After the chemicals are dissolved, adjust volume to 500 mL. Do not adjust pH.
3. Store reagent refrigerator. For best results aliquot into 50-100 mL bottles.

Gliadin and Glutenin Solvents
    The solvent used for the extraction of gliadins is 50% propan-1-ol (v/v). For glutenins, 50% propan-1-ol + 5% β-mercaptoethanol (v/v) is used. Other reducing agents such as dithiothreitol may also be used.

Apparatus

Several different companies now offer commercial HPCE instruments with a wide variety of configurations. Important features to consider are the presence and capacity of an autosampler, the available injection and detection modes, buffer vial capacity, the ease of using the instrument, the ease of use of the controlling software, and data analysis features, the ability to easily export data from the instrument, and the ability to thermostat the capillary. Several different approaches have been taken to control the temperature of the capillary, included liquid cooling, static air cooling and forced air cooling (Wehr et al 1999). This is usually viewed from the standpoint of cooling the capillary as excessive heat generation during a separation can result in poor resolution and reproducibility (Wehr et al 1999). Liquid cooling has been shown to remove more heat during a separation than forced air cooling, although forced air is a simpler system to use (Wehr et al 1999).

While it is important to remove heat generated during a separation, it is also important to be able to heat the capillary for some applications, this is especially true for some buffer systems for separating wheat proteins where increased run temperatures result in better separations (Lookhart and Bean 1995a).

The type of vials used for samples should also be considered. Some means of covering the samples to prevent evaporation of the sample matrix should be present. It is also desirable to be able to humidify the sample vials during the analysis (Bean and Lookhart 2001a). Sample vials for limited amounts of sample should also be available.

Procedure

The procedure outlined in this chapter utilizes the method reported by Bean and Lookhart (2000a) which is based on the isoelectric buffer, iminodiacetic acid (IDA) (Righetti and Bossi 1997). Isoelectric buffers are unique zwitterionic compounds that buffer without the need for co-ions (Capelli et al 1998). Due to the unique properties of these compounds, extremely high voltages can be used, resulting in very fast separations (Capelli et al 1998). These buffers were first reported for separating cereal proteins by Capelli et al (1998) and Righetti et al (1998). This group has published extensively on these buffering compounds including works on their theory (Righetti et al 1997) and practical considerations of their use (Bossi et al 1999). While the method presented here produces high resolution rapid separations, other successful FZCE methods have been developed for separating gluten proteins. Several reviews are available on the separation of cereal proteins by HPCE (Bean et al 1998, Bean and Lookhart 2000b, Bean and Lookhart 2001b). Note also that this procedure was optimized using Beckman PACE 2100 and 5510 instruments. Other instruments using different pressures, capillary thermostating methods, etc may need adjustments to this method for optimum results.

Sample Preparation

Gliadins

Gliadins can be prepared in a number of different ways and several solvents can be successfully used with FZCE (Bean and Lookhart 2001a). The following procedure is recommended for best reproducibility. A 10:1 ratio of solvent to flour is used, typically with 100 mg of flour and 1 mL of solvent (Bean and Lookhart 2001a).

1. Extract 100 mg flour with 1 mL of albumin and globulin extraction solvent (50 mM TrisHCl buffer, pH 7.8 containing 100 mM KCl and 5 mM EDTA) for 5 min with continual vortexing to remove albumin and globulin proteins (Bean and Lookhart 1998). To facilitate continual vortexing, samples are placed on a VortexGenie2 fitted with a 30 place foam vial holder.
2. Centrifuge and discard liquid.
3. Repeat steps 1 and 2.
4. Extract for 5 min (with continual vortexing) with deionized water to remove residual salt.
5. After centrifugation and removal of the water, add 1 mL of 50% propan-1-ol to the pellet and mechanical break up pellet with a spatula. Extracted for 5 min with continual vortexing.
6. After centrifugation, 0.5 mL of liquid is carefully removed and transferred to a new microfuge vial.
7. The remaining liquid is removed and the extraction with 50% propan-1-ol is repeated.
8. After centrifuging the second 50% propan-1-ol extract, 0.5 mL of liquid is again removed and added to the vial containing the first propan-1-ol extract (Bean and Lookhart 1998).
9. This liquid is then centrifuged to clarify and can be used directly for analysis. Sample filtration is generally not necessary.

Note that both whole meal and flour may be used to prepare samples for FZCE analysis of gliadins. However extracts prepared from flour produce better results and are more reproducible. Samples prepared from whole meal show less stability and capillary lifetimes are reduced when using samples prepared from whole meal. Likewise, it is possible to analyze gliadins without first extracting albumins and globulins. However, the presence of albumins and globulins will result in shorter capillary lifetimes.

Glutenins

To prepare samples for FZCE analysis, gliadins must be pre-extracted as they co-migrate with the glutenins (Lookhart and Bean 1995b, Bean and Lookhart 1997). The following procedure has been used to successfully prepare glutenins for FZCE analysis.

1. Extract 100 mg flour with 1 mL of 50% propan-1-ol for 5 min each with continual vortexing to remove ablumins, globulins, and gliadins (Bean *et al* 1998b). Centrigue and discard supernatant.
2. Repeat step 1 two additional times. Note some polymeric glutenin is also removed during these extractractions (Bean *et al* 1998b)
3. Add 1 mL of of 50% propan-1-ol containing 1% DTT or 5% BME and mechanical break up the pellet with a spatula. Extract for 30 min with continual vortexing.
4. Centrifuge and carefully remove the liquid to a new microfuge tube for analysis. The sample can be centrifuged to clarify before analysis, but filtration is not generally necessary.

Alkylation of the glutenins is not necessary and has been found to reduce resolution (Bean and Lookhart 1997) possibly because the conductivity of the sample plug is increased by the alkylating reagent and also possibly due to alteration of the charge on the proteins.

In addition to the "total" glutenin extracts mentioned above, both the high and low molecular weight glutenin subunits (HMW-GS and LMW-GS respectively) may be analyzed by FZCE using the methods described in this chapter. This may be desirable when analyzing glutenins by FZCE as the HMW-GS and LMW-GS show some overlap in their migration (Bean and Lookhart 1997) making it difficult to identify HMW-GS in a total glutenin analysis by FZCE. To prepare HMW-GS and LMW-GS acetone precipitation can be used as described in Melas *et al* (1994). Precipitation should be carefully carried out and samples should probably be analyzed by either SDS-PAGE or SDS-CE to verify the purity of the precipitations. This is especially important due to the overlap in the migration times of the glutenin subunit classes.

Capillary Preparation and Conditioning

One critical step in obtaining high resolution separations with HPCE is the proper preparation and condition of the capillary prior to the separations. Capillaries must be cut to specific lengths to be installed in HPCE instruments and the cut on the capillary ends can have a large impact on subsequent separations (reviewed in Rozing 1998). Therefore, clean flat cuts on the capillary ends are necessary for optimum resolution (Fig. 2).

The capillaries must be properly conditioned for optimum results. For wheat proteins several different rinsing protocols have been followed to prepare capillaries for separation (Bean *et al* 1998, Bean and Lookhart 2000b, Bean and Lookhart 2001b). However, a systematic investigation of the factors necessary to achieve good conditioning revealed that the most important factor was to rinse the capillaries with HPMC and to otherwise leave the inner walls of the capillaries alone (Bean and Lookhart 1998). Thus the most effective conditioning program is to rinse the capillary with separation buffer. This dynamically coats the inner walls with HPMC and also equilibrates the capillary to the separation pH. No other rinses are necessary and are can be detrimental to the equilibration of the capillary surface.

Thus, for a new capillary, or before a capillary is used for the day, it is simply flushed with separation buffer for 15 min. Between separations the capillaries are rinsed with separation buffer for 0.5 min. Capillaries can be stored overnight filled with buffer provided that the capillary ends are also covered with buffer to prevent evaporation and subsequent plugging of the capillaries. For longer term storage, capillaries can be rinsed with water and then nitrogen and stored dried.

Separation Conditions

Optimum separation conditions for the IDA buffer system have been established at 30 kV (1111 V/cm) and 45 °C (Bean and Lookhart 2000a). The separation voltage should be ramped up slowly as this gives higher efficient separations, probably due to less heating of the sample plug (Bean and Lookhart 2001a). Sample injection is carried out by pressure. The amount of sample injected needs to be optimized for each instrument with sample concentration balanced with the size of the sample plug injected (Bean and Lookhart 2001a). Many instruments are capable of injecting samples with voltage, however this will bias the sample towards proteins with higher mobilities (Wehr *et al* 1999).

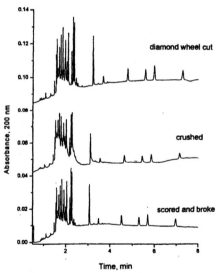

Fig. 2. Impact of capillary end-cut on resolution of gliadin separation by FZCE. From Bean and Lookhart (2001a) with permission.

For instruments injecting at 0.5psi, optimum injection conditions have been found to be 4 sec injections using samples extracted at a 10: 1 solvent to flour ratio (Bean and Lookhart 2001). The capillaries used by in these separations are 27 cm × 50 :m i.d. The 50 :m i.d. capillaries provide better sensitivity and do not plug as easily as some of the smaller 20 :m i.d. capillaries first used for wheat protein separations (e.g. Lookhart and Bean 1995a). Detector rise time should be set to a fast speed so that resolution is not compromised by the detector (Bean and Lookhart 2000a).

Reproducibility
Several factors can contribute to the success for failure of FZCE separations. As mentioned previously, capillaries must be fully conditioned and carefully prepared prior to use. Bean and Lookhart (1998, 2001a) investigated several factors that could influence the reproducibility of FZCE separations of wheat proteins. Those found to have an impact were sample vial humidification, capillary end-cut, capillary equilibration, sample injection amount, voltage ramp-up time, and sample preparation. Once these factors are properly addressed, FZCE is capable of good reproducibility (Fig. 3).

Another important aspect of maintaining high reproducibility is the proper monitoring of system performance. This includes monitoring the quality of the capillary. Proteins and other material can build up on the inner capillary walls over time, reducing the resolution of the separations. While certain additives, such as HPMC, can delay this process, none have been found to be 100% effective (Righetti paper). Thus it is important to monitor capillaries to determine when it is appropriate to replace them. One way to do this is to monitor the plates/meter of a given marker peak in a standard sample (Bean and Lookhart 2001). Over time, separation efficiency (plates/m) will decrease as capillaries start to fail. A standard sample should also be analyzed with each new capillary and with each data set to assess the resolution of the capillary. Often baseline resolution will start to decrease as a capillary starts to fail.

Another aspect that can be monitored is the current profile. Capillaries that start to fail sometimes show erratic currents. By monitoring the initial current of separation, the buffer quality can also be monitored.

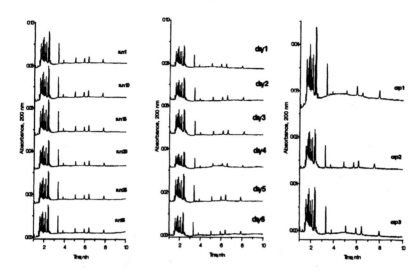

Fig. 3. Reproducibility of FZCE separations of wheat gliadins. Figures represent run-to-run, day-to-day, and capillary-to-capillary reproducibility, respectively. From Bean and Lookhart (2000a) with permission.

Examples of Applications

An example of gliadin separations from hard red winter (HRW) winter wheats are shown in Fig. 4. FZCE can be used to "fingerprint" cultivars providing a means of differentiation. The four cultivars shown in Fig. 4 are easily separated from each other. The gliadin subclasses have been located in FZCE separations and have been found to migrate in the same order as in A- PAGE (Lookhart and Bean 1995b, Bean and Lookhart 1997).

In addition to fingerprinting cultivars, FZCE has also been used to identify and differentiate between cultivars with 1BL.RS and 1AL.RS rye translocations (Lookhart *et al* 1996), to study proteins in transgenic wheats (Vasil *et al* 2001) and to monitor the purity of gliadin subclasses collected from preparative A-PAGE separations (Rumbo *et al* 1999). Extensive reviews of the applications of FZCE to the separation of wheat proteins are available (Bean *et al* 1998a, Bean and Lookhart 2000b, Bean and Lookhart 2001b, Bean and Lookhart 2001c). In addition to separating gliadins, both total glutenins and HMW-GS and LMW-GS can be successfully separated by FZCE. Fig. 5 shows an example of total glutenin and HMW-GS. It is important to note that the HMW-GS and LMW-GS co-migrate and that the proteins are separated by charge and not by size. This makes it more difficult to identify HMW-GS when separating the HMW-GS by FZCE.

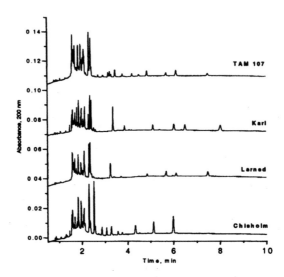

Fig. 4. Separation of gliadins from four HRW cultivars by FZCE. From Bean and Lookhart (2000a) with permission.

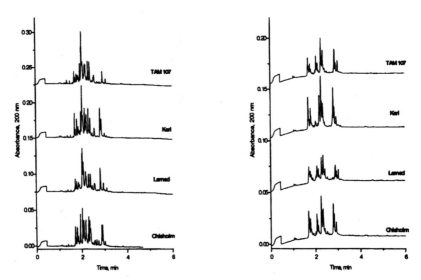

Fig. 5. Separtion of total glutenins (left) and HMW-GS (right) by FZCE. From Bean and Lookhart (2000a) with permission.

100

However, from 2D RP-HPLC × FZCE separations it appears that the HMW-GS migrate in the following order (from fastest mobility to slowest:) Dy, By, Bx/Ax, Dx (Bean and Lookhart 1997). Thus by using HMW-GS isolated from LMW-GS by precipitation it is possible to obtain some information on the HMW-GS. This is confounded however, by the fact that the HMW-GS are resolved into multiple peaks by FZCE (Bean and Lookhart 1997).

As mentioned above, FZCE can also be combined with RP-HPLC to produce 2D separations (Bean and Lookhart 1997). This can be done without the need for modification of the instruments using only a fraction collector (Bean and Lookhart 1997). Examples of 2D separations are shown in Fig. 6.

Tips and Modifications
While the FZCE procedure outlined in this chapter is capable of extremely high resolution rapid separations in (4 - 8 min), other buffer systems have successfully separated gluten proteins. These are fully reviewed in the following papers (Bean *et al* 1998, Bean and Lookhart 2000b, Bean and Lookhart 2001b, Bean and Lookhart 2001c). Since these buffer systems all differ somewhat, it would be possible to obtain different selectivities by changing buffer systems.

It is also possible to alter the separation conditions to produced different separations. For example, increasing the buffer concentration will lead to increased resolution but also slower separations. Furthermore, increasing the buffer concentration will lead to increased current production and could cause excessive Joule heating. An example of the effects of increasing buffer concentration on cereal proteins can be found in Lookhart *et al* (1999). Altering the separation temperature and voltage can also change the resolution of a separation. Likewise, several different organic solvents have been used to alter the resolution of FZCE separations (Lookhart and Bean 1995a, Bean and Lookhart 2000a) and could be used to vary the selectivity of separations. Capillary length may also be varied as can capillary diameter. Buffer pH can also be manipulated to produce separations with different selectivities. Beitz (1994) and Beitz and Schmalzried (1995) utilized a borate based buffer with a pH of 9.0 containing acetonitrile and SDS as a buffer additives. This buffer produces separations with much different patterns than acidic buffers. A charge reversal system has also been used to successfully separate wheat proteins (Werner *et al* 1994). In this system, the inner walls of the capillary are coated with a positively charged reagent and the proteins are separated in acidic buffer (Werner *et al* 1994). This is different from other acidic buffer render them neutral or inactive. The charge reversal system produces much different patterns than other acidic buffers and could be used to provide different selectivities.

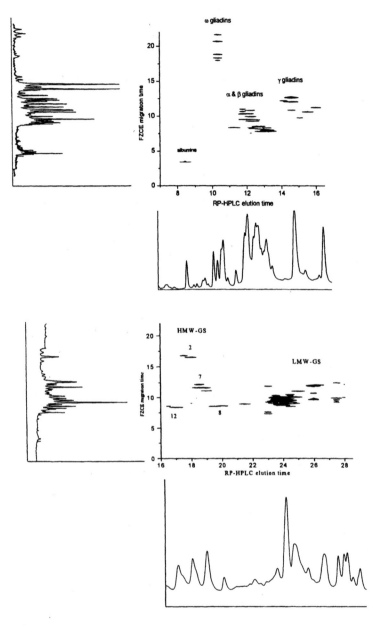

Fig. 6. Example of 2D RP-HPLC × FZCE separations of gliadins (top) and glutenins (bottom). Gliadin subclasses are labeled on the gliadin separations, while HMW-GS are numbered on the glutenin separations. From Bean and Lookhart (1997) with permission.

Principles

While FZCE produces separations that may be consider analogous to A-PAGE, sodium dodecyl-sulfate capillary electrophoresis (SDS-CE) produces separations analogous to SDS-PAGE, i.e. size based separations. In fact, the first SDS-CE separations simply utilized capillaries with polyacrylamide gels polymerized inside the capillaries (Cohen and Karger 1987). Problems with this method led to new methods using replaceable entangled polymer solutions (Wehr *et al* 1999). Unlike fixed gels, these polymer solutions can be pumped through the capillary after each separation. Protein-SDS complexes are separated by size or "sieved" as the move through the entangle polymer matrix inside the capillary. The polymer matrix can then be emptied from the capillary and fresh matrix put in place for the next separation. An diagram of an SDS-CE separation of gluten proteins is shown in Fig. 7.

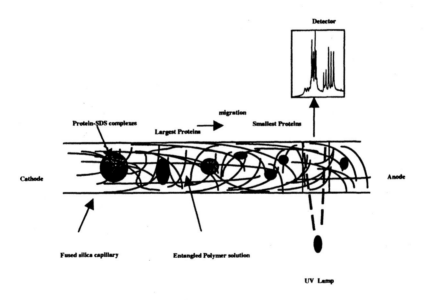

Fig. 7. Diagram of an SDS-CE separation of gluten proteins.

Several different types of polymers have been successfully used to produce SDS-CE separations of proteins (Wehr *et al* 1999). Several have been used to separate gluten proteins including uncross-linked polyacrylamide (Werner *et al* 1994, Bean and Lookhart 1999), dextran (Bean and Lookhart 1999, polyethylene oxide (PEO) (Bean and Lookhart 1999, Zhu and Khan 2001), and a commercial reagent from BioRad (Bean and Lookhart 1999). SDS-CE separations are typically carried out at neutral or basic pH where the capillary walls are negatively charged. This presents some problems as the electroendosmotic flow (EOF) increases as pH increases (Wehr *et al* 1999). EOF is the bulk flow of liquid inside the capillary, which is caused by charges on the inner surface of the capillaries (Wehr *et al* 1999). EOF is generally considered detrimental to HPCE separations because it is difficult to reproduce the magnitude of the flow between separations. To avoid the problems of EOF in SDS-CE separations, permanently coated capillaries are often used (e.g. Ganzler *et al* 1992). However, polymers which bind to the capillary walls and reduce EOF have also been used in SDS-CE systems to separate gluten proteins. These have included uncross-linked polyacrylamide (Werner *et al* 1994), PEO (Bean and Lookhart 1999), and commercial reagents (Bean and Lookhart 1999). Extensive, more complete reviews on SDS-CE are available in the literature (Heller 1995, Guttman 1996, Chrambach 1996, Takagi 1997).

Reagents

All reagents used for SDS-CE separations should be of the highest possible purity (e.g. Simga Ultrapure). Impurities in reagents used in buffer preparation can cause increased currents, may bind to capillary walls, or possibly interact with proteins themselves. When filtering reagents containing SDS, large batches should be filtered at one time as significant variation in the amount of SDS bound to different filters was found when filtering smaller batches and using fresh filters for each batch (Kelly *et al* 1997). Due to the viscosity of the polymer solutions used in SDS-CE, the solutions may need to be degassed prior to use to avoid introducing air bubbles into the capillary. This is easily accomplished by centrifuging the solutions before use or by brief sonication.

Separation Buffer

As this procedure utilizes a commercial buffer, little preparation is needed. The commercial SDS-CE reagent from BioRad is mixed with ethylene glycol (EG) to give a final concentration of 15% EG (Bean and Lookhart 1999). Note that both these reagents are viscous and careful measuring and mixing of the reagents is necessary. Mixing of the reagents also introduces air bubbles into the mixture so degassing is necessary (easily accomplished by centrifuging the mixture or by letting it stand overnight). Note that we have been told conflicting information about the continued availability of this product. New polymer systems are currently being investigated.

Apparatus

Several different companies now offer commercial HPCE instruments with a wide variety of configurations. Important features to consider are the presence of an autosampler, the capacity of the autosampler, the available injection and detection modes, buffer vial capacity, the ease of using the instrument, ease of use of the controlling software, ease of use of data analysis features, ability to easily export data from the instrument, and the ability to control the temperature of the capillary. Several different approaches have taken to control the temperature of the capillary, included liquid cooling, static air cooling and forced air cooling (Wehr *et al* 1999). This is usually viewed from the standpoint of cooling the capillary as excessive heat generation during a separation can result in poor resolution and reproducibility (Wehr *et al* 1999). Liquid cooling has been shown to remove more heat during a separation than forced air cooling, although using forced air cooling may make changing capillaries simpler (Wehr *et al* 1999).

The pressure rinsing capabilities of an instrument should also be considered when performing SDS-CE. Instruments with higher pressure rinses are able to fill capillaries with more viscous polymer solutions, which might be advantageous for some applications.

While it is important to remove heat generated during a separation, it is also important to be able to heat the capillary for some applications, this is especially true for some buffer systems for separating wheat proteins where increased run temperatures result in better separations (Lookhart and Bean 1995a). Higher temperatures have also been found to influence different types of polymer matrices in different ways (Guttman *et al* 1993).

Procedure

The procedure follows that reported in Bean and Lookhart (1999) using a commercial SDS-CE reagent. As with FZCE, several SDS-CE polymer systems and methods have been successfully used to separate wheat proteins. Several reviews on the SDS-CE separations of wheat proteins are available (Bean *et al* 1998a, Bean and Lookhart 2000b, Bean and Lookhart 2001b, Bean and Lookhart 2001c).

Sample Preparation

Gliadins
1. 100 mg flour are extracted with 1 mL of 1% SDS for 5 min with continual vortexing (Bean and Lookhart 1999).

105

2. The mixture is centrifuged and the liquid transferred to a new microfuge tube and heated at 100 °C in a heating block for 5 min and is then centrifuged again.
3. After centrifugation the liquid can be used directly for SDS-CE analysis.

Samples do not need to be filtered. Note that albumins and globulins can be pre-extracted as described in the FZCE section (see above) before extraction of the gliadins if desired.

Glutenins
1. 100 mg samples are first extracted with 1 mL of 50% propan-1-ol to remove gliadins and albumins and globulins (Bean *et al* 1998 b).
2. Centrifuge and discard liquid.
3. Repeat steps 1 and 2 two times. Note that some soluble polymeric protein will also be solublized during these steps.
4. Add 1 mL of 1% SDS + 5% BME to the pellet and mechanical mix with a spatula. Extract for 30 min (with continual vortexing).
5. Centrifuge the sample and carefully transfer the liquid to a new microfuge tube.
6. Heat sample at 100°C for 5 min in a heating block. This liquid is centrifuged again and then can be used directly for analysis. Samples do not need to be filtered before injection.

HMW-GS can also be isolated by using the sequential extraction procedure and precipitation by acetone as described for FZCE (see above). Precipitated HMW-GS should be redissolved in 1% SDS/5% BME and heated as described for gliadins and glutenins.

Note that all these sample handling procedures call for the use of unbuffered 1% SDS solutions to be used as the sample matrix. Buffer salts increase the conductivity of the sample plug which can cause sample stacking problems. Sutton and Bietz (1997) found that unbuffered sample matrix provided better baselines and more reproducible results for wheat proteins than sample matrix's containing buffer salts.

Capillary Preparation and Equilibration
When preparing the capillary for initial use care must be taken to insure a clean square cut on the capillary ends (see Fig. 2 above). The first time a capillary is used (initially or for the day) it is rinsed for 5 min with 1M NaOH followed by 5 min 1M H Cl. The capillaries are then rinsed with the BioRad SDS-CE/EG mixture for 15 min. Prior to each separation the capillaries are rinsed with the BioRad SDS-CE/EG mixture for 5 min. After each separation, the capillaries are rinsed with 1M NaOH for 2 min followed by 1M H Cl for 1 min. Capillaries can be rinsed with water and then nitrogen and stored dry when not in use. For overnight storage, capillaries can be rinsed with separation buffer and stored with the ends of

the capillary covered by separation buffer. This reduces the time necessary to equilibrate the capillaries the following morning, but caution should be used to insure that the capillary ends are covered so that buffer cannot evaporate and plug the capillaries.

Separation Conditions

Samples are separated at 8kV and 30°C (Bean and Lookhart 1999). The voltage should be ramped up slowly to improve the separation efficiency (Bean and Lookhart 2001a). Samples are injected with pressure for 15-45 sec depending on sample concentration. Sample times need to be optimized and balanced with sample concentration for optimum reproducibility (Bean and Lookhart 2001a). Voltage injection should be avoided if possible to prevent biasing the sample amounts injected.

Reproducibility

As discussed in the section on FZCE (see above) several factors can influence the reproducibility of HPCE separations. Standards should be analyzed with each data set to insure proper functioning of the system and quality of the capillary. With SDS-CE systems, polymer can build up on the electrodes of the instruments and influence the reproducibility. Keeping the instruments clean helps to improve the overall reproducibility and reliability of the results (Wehr *et al* 1999). The overall reproducibility of SDS-CE systems for wheat proteins has been reported to be very good (Sutton and Bietz 1997, Bean and Lookhart 1999).

Examples of Applications

Several different fractions of gluten proteins have been separated by SDS-CE. Fig. 8 shows the separation of total protein, gliadin, and glutenins in the SDS-CE system presented here. Note that some late migrating gliadins, probably T-gliadins co-migrate with the fastest moving HMW-GS. This should be taken into account when trying to quantitate the amount of HMW-GS in total protein extracts using this SDS-CE system.

SDS-CE can also be used to differentiate cultivars based on gliadin separations. Fig. 9 shows separation of gliadins from five different wheat cultivars. Note that each cultivar shows a distinct profile, although the resolution is much lower than that obtained by FZCE for gliadins.

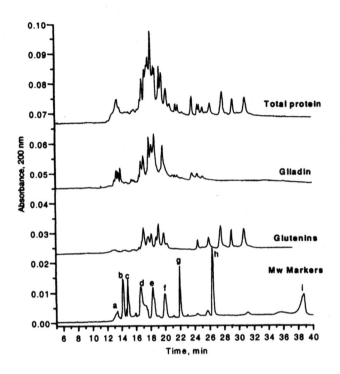

Fig. 8. Separation of wheat protein fractions in BioRad SDS-CE/EG system. From Bean and Lookhart (1999) with permission.

Most papers on the separation of gluten proteins by SDS-CE have focused on the HMW-GS, either as part of the glutenin fraction or prepared by selective precipitation. Separation of the HMW-GS using the BioRad/EG SDS-CE system is shown Fig. 10. Profiles of the same samples separated by SDS-PAGE are also shown for comparison. As can be seen in Fig. 10, the migration order of the HMW-GS in SDS-CE varies from that in SDS-PAGE. This has been reported in several papers on SDS-CE (reviewed in Bean *et al* 1998a). Furthermore, the migration order appears to vary in different SDS-CE systems (Zhu and Khan 2001). Thus caution should be used when assigning HMW-GS composition using SDS-CE.

Figure 10 also shows that SDS-CE resolved all the allelic forms of the HMW-GS from each other in each cultivar. LMW-GS were also separated with high resolution. Zhu and Khan (2001) have reported the use of SDS-CE for the quantitation of HMW-GS using a PEO separation matrix.

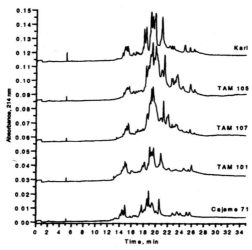

Fig. 9. Separation of gliadins from five different wheat cultivars in the BioRad/EG SDS-CE system. From Bean and Lookhart (1999) with permission.

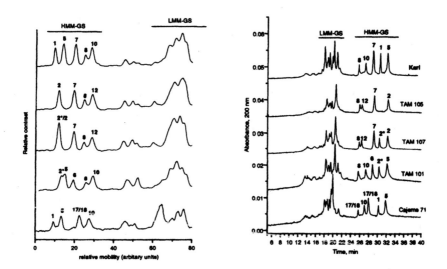

Fig. 10. Separation of glutenins from different wheat cultivars by SDS-CE (left) and SDS-PAGE (right). Lane profiles for SDS-PAGE were generated from a digital camera using image analysis software (Kodak 1D gel analysis software).

Tips and Modifications

Again as with FZCE, several factors in SDS-CE separations can be varied to modify the separations. Buffer concentration, buffer pH, separation voltage, separation temperature, polymer type, polymer Mr, polymer concentration, buffer additives, and capillary length could all potentially be varied to change separations for specific applications.

Most of the work on gluten proteins separated by SDS-CE has focused on two areas, polymer type and buffer additives. Several different polymer systems have been reported for the separation of wheat proteins. Bean and Lookhart (1999) showed different patterns for gluten proteins when four different polymers were used to separate the proteins. For example uncross-linked polyacrylamide showed lower resolution of the LMW-GS and gliadins than the other polymers tested, but good resolution of the HMW-GS (Bean and Lookhart 1999). PEO showed much different patterns than the other polymers (Bean and Lookhart 1999) and also showed different migration orders for the HMW-GS than some of the other polymer systems (Zhu and Khan 2001).

The addition of rganic solvents to the separation buffer can improve the resolution of the HMW-GS, however, in some buffer systems such as PEO this is not the case. The type of additive used has also been found to vary with the polymer used (Werner *et al* 1994, Bean and Lookhart 1999). It may be possible, therefore, to vary these parameters to improve resolution of specific fractions of gluten proteins. The development of SDS-CE for the separation of gluten proteins has been reviewed by Bean *et al* (1998a).

Conclusions

Considerable work has been done to develop methods for the separation of gluten proteins (and other cereal proteins) by HPCE. The methods presented here have been reported to work with good reproducibility. References to many other methods have also been listed here to provide the reader with flexibility to analyze gluten proteins by several different methods. The ability to easily change separation conditions is one advantage of HPCE.

In addition to methods for cereal proteins, several papers on the general use of HPCE may be useful to those wishing to use HPCE to study gluten proteins. These include, Altria and Campi (1999a, b), Hows *et al* (1997), Altria (1999), Whatley (1999), Altria *et al* (1997), Altria and Hindocha (1998), Lucy *et al* (1998), Chapman and Hobbs (1999), Heiger *et al* (1997), and Corradini and Cannarsa (1996). These articles provide practical information on the use of HPCE and together with this chapter, should provide the information necessary to successfully separate gluten proteins by HPCE.

References

Altria, K.D., Bryant, S.M., Clark, B.J., and Kelly, M.A. 1997. The care and maintenance of CE capillaries. LC-GC 15:34-38.

Altria, K.D., and Hindocha, D. 1998. Instrument issues: operational qualification and method transfer. LC-GC 16: 835-841.

Altria, K.D. 1999. Optimization of sensitivity in capillary electrophoresis. 17:28-25.

Altria, K.D, and Campi, F. 1999a. Ten ways to spoil a CE separation. LC-GC 17:236-243.

Altria K.D., and Campi, F. 1999b. Another 10 ways to spoil a CE separation. LC-GC 17:828-834.

Bean, S.R. and Lookhart, G.L. 1997. Separation of wheat proteins by two-dimensional reversed-phase high-performance liquid chromatography plus free zone capillary electrophoresis. Cereal Chem. 74:758-765.

Bean, S.R. and Lookhart, G.L. 1998. Faster capillary electrophoresis separations of wheat proteins through modification to buffer composition and sample handling. Electrophoresis. 19, 3190-3198.

Bean, S.R., Bietz, J.A., and Lookhart, G.L. 1998a. High performance capillary electrophoresis of cereal proteins. J. Chrom. A. 814:25-41.

Bean, S.R., Lyne, R.K., Tilley, K.A., Chung, O.K., and Lookhart, G.L. 1998b. A rapid method for quantization of insoluble polymeric proteins in flour. Cereal Chem. 75:374-379.

Bean, S.R., Lookhart, G.L. 1999. SDS-CE separations of wheat proteins. I. Uncoated capillaries. J. Ag. Fd. Chem. 47, 4246-4255.

Bean, S.R. and Lookhart, G.L. 2000a. Ultrafast capillary electrophoretic analysis of cereal storage proteins and its applications to protein characterization and cultivar differentiation. J. Ag and Fd. Chem. 48, 344-353.

Bean, S.R. and Lookhart, G.L. 2000b. Electrophoresis of cereal storage proteins. J. Chrom. A. 881: 23-36.

Bean, S.R., and Lookhart, G.L. 2001a. Optimizing quantitative reproducibility in HPCE separations of cereal proteins. Cereal Chem. 78:530-537.

Bean, S.R. and Lookhart, G.L 2001b. Recent advances in the HPCE of cereal proteins. Electrophoresis. 22:1503-1509.

Bean, S.R., and Lookhart, G.L. 2001c. HPCE of meat, dairy, and cereal proteins. Electrophoresis 22:4207-4215.

Bietz, J.A., and Schmalzried, E. 1995. Capillary electrophoresis of wheat gliadin: Initial studies and application to varietal identification. Lebensmittel-Wissenschaft und-Technologie 28:174- 184.

Bietz, J.A. 1994 in Proc. 5th Int. Workshop on Gluten Proteins, Assoc. Cereal Res., Detmold, Germany.

Bossi, A., Olivieri, E., Castelletti, L., Gelfi, C., Hamdan, M., and Righetti, P.G. 1999. General experimental aspects of the use of isoelectric buffers in capillary electrophoresis. J. Chromatog. A 853:71-82.

Bossi, A., and Righetti, P.G. 1997. Generation of peptide maps by capillary zone electrophoresis in isoelectric iminodiacetic acid. Electrophoresis, 18: 2012-2018.

Capelli, L., Forlani, F., Perini, F., Guerrieri, N., Cerletti, P., and Righetti, P.G. 1998. Wheat cultivar discrimination by capillary electrophoresis of gliadins in isoelectric buffers. Electrophoresis. 19: 311-319.

Chapman, J. C., and Hobbs, J. 1999. Putting capillary electrophoresis to work. LC-GC 17:86-99.

Chrambach, A. 1996. Quantitative and automated electrophoresis in sieving media. Electrophoresis 17: 454-464.

Cohen, A.S., and Karger, B.L. 1987. High-performance sodium dodecyl sulfate polyacrylamide gel capillary electrophoresis of peptides and proteins.J Chromatogr 1987 Jun 26;397:409-17.

Corradini, D., and Cannarsa, G. 1996. Capillary electrophoresis analysis of proteins in bare fused-silica capillaries. LC-GC 14:326-332.

Ganzler, K., Greve, K.S., Cohen, A.S., Karger, B.L. 1992. High-performance capillary electrophoresis of SDS-protein complexes using UV-transparent polymer networks. Anal. Chem. 64: 2665-2671.

Guttman, A., Horvath, J., and Cooke, N. I. 1993. Influence of Temperature on the Sieving Effect of Different Polymer Matrices in Capillary SDS Gel Electrophoresis of Proteins. Anal. Chem. 65: 199-203.

Guttman, A. 1996. Capillary sodium dodecyl sulfate-gel electrophoresis of proteins. Electrophoresis 17: 1333-1341.

Guzman, N.A., ed. 1993. Capillary Electrophoresis Technology. Marcel Dekker, New York.

Hegarty, V. 1995. Proteins-meat, fish, and other protein sources. Pages 197-226 in: Nutrition, Food, and the Environment. Eagen Press: St. Paul.

Heiger, D. Majors, R.E., and Lombardi, R.A. 1997. Method development strategies in capillary electrophoresis. LC-GC 15:14-23.

Heller, C. 1995. Capillary electrophoresis of proteins and nucleic acids in gels and entangled polymer solutions. J. Chromatogr. A. 698: 19-31.

Hows, M.E.P., Alfazema, L.N., and Perret, D. 1997. Capillary electrophoresis buffers-approaches to improving their performance. LC-GC 15:1034-1043.

Kelly M.A., Altria, K.D., and Clark B.J., 1997. Quantitative Analysis of Sodium Dodecyl Sulphate by Capillary Electrophoresis. J.Chromatogr.A, 781:67-71

Landers, J.P., ed. 1997. Handbook of Capillary Electrophoresis, 2nd ed., CRC Press, Florida.

Lookhart, G.L, and Bean, S.R. 1995a. A fast method for wheat cultivar differentiation using capillary zone electrophoresis. Cereal Chem. 72:42-47.

Lookhart, G. and Bean, S.R. 1995b. Separation and characterization of wheat protein fractions by high-performance capillary electrophoresis. Cereal Chem. 72:527-532.

Lookhart, G.L., Bean, S.R., Graybosch, R, Chung, O.K., Morena-Sevilla, B., and Baenziger, S. 1996. Identification by high-performance capillary electrophoresis of wheat lines containing the 1AL.1RS and the 1BL.1RS translocation. Cereal Chem. 73:547-550.

Lookhart, G.L., Bean S.R., and Jones, B.L. 1999. Separation and characterization of barley (Hordeum vulgare L.) hordeins by free zone capillary electrophoresis. Electrophoresis. 20, 1605-1612.

Lucy, C. A., Yeung, K. K.-C., Peng, X., and Chen, D.D.Y. 1998. Extraseparation peak broadening in capillary electrophoresis. LC-GC 16: 26-32.

MacRitchie, F. 1992. Wheat Proteins PhysicoChemistry and Functionality Pages 1-87 in Advances in Food and Nutrition Research. J. E. Kinsella, ed., Academic Press, Inc.: San Diego.

Melas, V., Morel, M.-H., Autran, J.-C., and Feillet, P. 1994. Simple and Rapid Method for Purifying Low Molecular Weight Subunits of Glutenin from Wheat. Cereal Chem. 71:234-237.

Righetti, P.G., ed. 1996. Capillary Electrophoresis in Analytical Biotechnology. CRC Press, Florida.

112

Righetti, P.G., Gelfi, C., Perego, M., Stoyanov, A.V., and Bossi, A. 1997. Capillary zone electrophoresis of oligonucleotides and peptides in isoelectric buffers: Theory and methodology. Electrophoresis, 18: 2145-2153.

Righetti, P.G., Oliveri, E., and Viotti, A. 1998. Identification of maize lines via capillary electrophoresis of zeins in isoelectric, acidic buffers. Electrophoresis 19: 1738-1741.

Rozing, G.P. 1998. Importance of a clean column end for CE and CEC capillaries. American Laboratory. Dec:33.

Rumbo M., Chirdo F.G., Giorgieri S.A, Fossati C.A., and Anon M.C. 1999. Preparative fractionation of gliadins by electrophoresis at pH 3.1 (A-PAGE). J Agric Food Chem 47:3243-7.

Sutton, K.H., and Bietz, J.A. 1997. Variation among high molecular weight subunits of glutenin detected by capillary electrophoresis. J. Cereal Sci., 25: 9-16.

Takagi, T. 1997. Capillary electrophoresis in presence of sodium dodecyl sulfate and a sieving medium. Electrophoresis 18: 2239-2242.

Vasil, I.K, Bean, S.R., Zhao, J., McCluskey, P., Lookhart, G.L., Zhao, H., Altpeter, F., and Vasil. V. 2001. Evaluation of baking properties and gluten protein composition of field grown transgenic wheat lines expressing high molecular weight glutenin gene 1Ax1. Journal of Plant Physiology 158: 521-528.

Wehr, T., Rodriguez-Diaz, R., and Zhu, M. 1999. Capillary Electrophoresis of Proteins. Marcel Dekkar, New York.

Werner, W.E., Wiktorowicz, J.E. and Kasarda, D.D. 1994. Wheat varietal identification by capillary electrophoresis of gliadins and high molecular weight glutenin subunits. Cereal Chem. 71: 397-402.

Whatley, W. 1999. Making CE work-points to consider. LC-GC 426-432.

Zhu, J., and Khan, K. 2001. Separation and quantification of HMW glutenin subunits by capillary electrophoresis. Cereal Chem. 78:737-742.

Chapter 6

Size Exclusion Chromatography and Flow Field-Flow Fractionation of Wheat Proteins

Ken R. Preston and Susan G. Stevenson
Grain Research Laboratory, Canadian Grain Commission, Winnipeg, MB,
R3C 3G8, Canada

Introduction

The size distribution of the gluten proteins, particularly the polymeric proteins, plays a primary role in the processing quality of wheat flour. Size exclusion chromatography (SEC) has been the most widely used technique for the size characterization of wheat proteins. More recently, flow field-flow fractionation (flow FFF) and multistacking SDS polyacrylamide gel electrophoresis (multistacking SDS-PAGE) have been used to extend the upper size measurement range relative to SEC. In this chapter, the principles and the methods and conditions for fractionation of wheat proteins by SEC and flow FFF are presented. Information on multistacking SDS-PAGE is presented in chapter 3.

Size Exclusion Chromatography

Principles

Size exclusion chromatography (SEC) is a family of techniques that use porous packing to fractionate macromolecules and/or complexes based on their hydrodynamic radius. Separation is based on the ability of a component to penetrate the pores of the packing. Components with smaller diameters can penetrate into smaller pores than larger diameter analytes. The smaller components thus spend a higher portion of their time in the pore volume relative to the interstitial volume resulting in longer elution times. Components with radii greater than the maximum pore size elute at the void (interstitial) volume while those with radii less than the minimum pore size elute at the total permeation volume (sum of pore and interstitial volume). Those of intermediate size elute at volumes based on the radius of the component and the pore-size distribution of the packing.

Neue (1997) provided an excellent review of the theory of SEC. If a spherical molecule of radius r_m is eluted through a packing material with cylindrical bottomless pores of uniform radius R_{po}, then the fraction of pore space in which the molecule can reside is given by

$$K_o = (1-r_m/R_{po})^2$$

where K_o is the partition coefficient. However, since macromolecules act more like random coils under most SEC conditions and there is a distribution of pore sizes, the partition coefficient can be expressed as

$$K_o = \int\int P_r P_p P_R dR_{po} dr_m$$

where P_r is the probability that a macromolecule forms a random coil of radius r, P_p is the probability that this random coil will reside in a pore with radius R and P_R is the distribution of pore radii.

Plots of K_o versus r_m (or elution volume) show steeper slopes when either K_o values approach 1 for larger macromolecules ($r_{m=>}R_{po}$) which elute near the interstitial volume or approach 0 for small macromolecules ($r_m<<R_{po}$) which elute near the total permeation volume. Resolution, which is inversely proportional to this slope, is therefore optimized when conditions (column type and mobile phase) are chosen such that components elute at K_o values that are not approaching 1 or 0.

A number of other factors may influence the partition coefficient (for reviews see Potschka 1988 and Neue 1997). Interactions can occur among components and between components and the packing material through ionic, hydrophobic and hydrophilic interactions. These effects should be minimized through the appropriate choice of packing material and mobile phase as discussed in later sections. The impact of the mobile phase on the shape (hydrodynamic radius) of the components can also influence elution characteristics leading to erroneous estimates of molecular size. Components larger than the pore exclusion limit can experience partitioning due to "hydrodynamic drag" near the surface of the packing material where the velocity of the mobile phase is reduced towards zero flow. The larger the component, the more it is excluded from this interstitial space. Therefore, very large components can elute at volumes smaller than the theoretical void volume based on interstitial volume. Temperature, in theory, should not influence the partition coefficient since the distribution of a component between the mobile and stationary phase is controlled by entropy. However, secondary temperature induced changes in the particle size of the packing material and the hydrodynamic volume of components with extended structures can influence results.

116

Methods and Conditions

Columns

Two types of SEC are generally recognized based on the response of the packing material to pressure during elution. SEC that uses packing material that requires very low pressure to avoid compression is normally termed gel filtration chromatography while packing material that can withstand higher pressures without compression is normally termed size exclusion high performance liquid chromatography or SE-HPLC.

SEC was originally developed using "soft gels" prepared from loosely woven strands of dextran polymers (Sephadex), beaded agarose (Sepharose), polyacrylamide gels (Biogel) or cross linked dextran (Sephacryl). Columns are normally packed manually and run at low flow rates using either gravity feed or syringe or peristaltic pumps to minimize pressure pulsing. The larger diameter of the gel material reduces resolution requiring the use of relatively large columns to obtain good resolution. This results in considerably longer run times relative to the smaller more efficient rigid packing materials used for SE-HPLC. For this reason, gel filtration chromatography of proteins is now used primarily for large scale preparation and desalting. In spite of these drawbacks, earlier gel filtration studies were useful for fractionation and characterization as well as assessing the functionality of wheat proteins. Beckwith *et al* (1966) separated gliadins into three fractions (HMW gliadin, ω-gliadin and LMW gliadin) while Huebner and Wall (1974) were able to fractionate reduced glutenin into three fractions using Sephadex G-100 or G-200. Huebner and Wall (1976) also used porous agarose (Sepharose 4B) to fractionate unreduced wheat flour proteins into two glutenin fractions as well as fractions corresponding to gliadins and water-soluble proteins. They were able to show that flour with longer mixing times and stronger dough properties had a higher ratio of glutenin in the high MW peak relative to the lower MW peak. Later studies using Sepharose CL-2B and Sephacryl S300 also indicated a strong relationship between dough strength and/or baking quality and the amount and size distribution of the glutenin proteins (Hamada *et al* 1982, Preston *et al* 1992). Studies with gels with higher exclusion limits such as Sepharose CL-4B also suggested that some glutenins had MW's extending into the millions or 10's of millions (Huebner and Wall 1980, Rao and Nigam 1988). Since gel filtration has now been almost completely replaced by SE-HPLC, procedures will not be discussed in this chapter. The reader is referred to the cited papers for methodology.

The introduction of derivatized silica packing with a range of average pore sizes by Toyo Soda in Japan in the late 1970's and early 1980's revolutionized SEC. These rigid packings are capable of withstanding pressures of 3000 psi or more allowing the use of small particle (about 5-15 μm) sizes that greatly enhance column efficiency (plate counts over 7,000

117

plates/m). Smaller columns at relatively high flow rates can thus be used resulting in high resolution with short run times. The use of these shorter silica columns, combined with efficient high pressure piston pumps and related equipment, allowed the development of SEC techniques that were initially coined as "size exclusion high pressure liquid chromatography" but are now generally referred to as size exclusion high performance liquid chromatography (SE-HPLC).

Two types of derivatized silica are widely available. Silica derivatized with tri-methylsilane groups is used for organic SE-HPLC while silica derivatized with hydrolyzed glycidoxypropylsilane (diol silica) is used for aqueous SE-HPLC, including proteins, where a hydrophilic particle surface is appropriate. Both types of derivatized silica contain some residual silanols, which can interact with ionic groups of the polymer and influence retention characteristics. For diol silica, these interactions can be virtually eliminated by the appropriate choice of mobile phase (pH adjustment and/or salt). There is a range of average pore sizes available for diol silica. TSK diol silicas from Toyo Soda have pore sizes of 250 Å (TSK-3000SW) and 450 Å (TSK-4000SW). For globular proteins, molecular weight fractionation ranges for these columns are about 1,000 to 300,000 and 4,000 to 1,000,000, respectively. However, for proteins with inherent or denaturing solvent (e. g. SDS) induced extended structures, the upper range can be considerably smaller due to their increased hydrodynamic radius. In recent years, silica supports with pore sizes of 500 Å have been developed by companies such as Shodex (Protein KW-804), which have upper fractionation ranges of about 2,000,000 for globular proteins. Smaller particle (5-7 μm) sizes for many newer silica supports also allow separations in less than 15 min. Drawbacks of silica supports include their instability at low (pH<2.5) and high (pH>7.5) pH and their brittle characteristics which limits pore size and thus resolution of very large components and stable complexes.

A number of more rigid organic polymers have been developed for SEC. Hydrophobic styrene-divinylbenzene co-polymers have been used for organic SEC while hydrophilic cross-linked glycidoxymethacrylates, cross-linked polyvinylalcohol and cross-linked agarose have been developed for aqueous SEC. These polymers can withstand pressures of about 200-400 psi, which, arguably, makes them suitable for SE-HPLC. The major potential advantages these supports have over silica are stability to a wider range of pH, very high column efficiency (> 30,000 plates/m) and their availability in larger pore sizes which increases the upper limit for fractionation. For example, polyhydroxymethacrylate packings from Shodex (OHpak SB-HQ) and from Agilent (PL aquagel-OH) are available with pore sizes of 1000 Å, allowing separation of globular proteins with molecular weights up to about 10,000,000. Amersham offers two cross linked dextrans, namely Sepharose 6 and Sepharose 12, which have reported protein exclusion limits of 4×10^7 and 2×10^6, respectively.

118

Smaller particle sizes (<15 μm) in these organic polymers allow high resolution with run times approaching that of silica based packing. As with silica packing, these organic packings also have some surface charges, which can lead to component retention unless an appropriate eluent is used.

Most SE-HPLC studies involving wheat proteins have used TSK-3000SW, TSK-4000SW or equivalent diol silica columns. Earlier work with these columns has been reviewed by Autran (1994). Derivatized silica columns continue to be dominant in more recent studies. Many of these studies show a trend towards the use of silica supports with the largest pore sizes available (up to 500 Å), which has improved the resolution of the higher molecular weight glutenin peaks (e. g. Bangur *et al* 1997, Nightingale *et al* 1999, Ammar *et al* 2000, Lindsay *et al* 2000, Carceller and Aussenac 2001, Bean and Lookhart 2001). Separose 6 and Sepharose 12 columns have also been used (e. g. Autran 1994, , Cornec *et al* 1994, Ciaffi *et al* 1996, Huebner *et al* 1997, Linarès *et al* 2000, Aussenac *et al* 2001). This can probably be attributed to their better resolution of larger glutenin peaks due to higher exclusion limits (Pasaribu *et al* 1992). Recently, Carcellar and Aussenac (2001) have used multiangle laser light scattering detection to demonstrate that PL-aquagel–OH 60 provides much superior resolution of the larger glutenin proteins than a silica based column (TSK G 4000 SW).

Mobile phase

There are two major requirements for the mobile phase for SEC (Neue 1997). The first is that it must be a good solvent for the components under study. For proteins, the solvent should reduce inter-protein interactions and maintain "native" conformation. The latter is an exception to the general rule for SEC of most polymers where a non-interacting random coil is preferable. Maintaining "native" structure is also an advantage in post run analysis where structural information is an objective.

The second major requirement for the mobile phase is the elimination of interactions between components and the stationary phase. For proteins, low concentrations of salt are normally added to aqueous solvents to suppress ionic interactions with residual ionic groups on the hydrophilic SEC stationary phase. Higher ionic strength should be avoided to prevent increased hydrophobic interactions. Methanol, acetonitrile or chaotropic agents such as SDS can also be added to reduce inter-protein and protein stationary phase hydrophobic interactions. Other factors that should be considered in the choice of mobile phase include compatibility with the extraction solvent (if different), impact on post column detection (e. g. low absorbance) and low viscosity to reduce back pressure and increase diffusion of components.

For SE-HPLC of wheat proteins, the most widely used mobile phases have been dilute sodium phosphate (or borate) buffers containing 0.1 or 0.05% SDS and acetonitrile/water (50:50) containing 0.1% or 0.05% TFA

(Autran 1994, Cornec *et al* 1994). These eluents show relatively low viscosity which reduces back pressure, have low absorbance at 210-220 nm and maintain protein extracts including the "more insoluble" glutenin in "solution". Higher SDS concentrations, other chaotropic agents (urea, DMF) and alcohol/water based solvents used earlier with gel filtration cause high back pressure and reduce column life (Batey *et al* 1991). High absorbance at 210-220 nm for some of these solvents can also be an issue. In general, there has been a trend towards the use of acetonitrile/water with TFA over SDS due to superior resolution and increased column life (Batey *et al* 1991). The acetonitrile based eluent is volatile and can be easily removed whereas SDS binds tightly to the protein and is difficult to remove for further characterization.

Extraction of wheat proteins for SE-HPLC analysis

Although characterization of the size distribution of specific wheat protein solubility fractions has proven useful (see review by Autran 1994), the emphasis of most SE-HPLC studies has been directed towards analysis of the overall protein profiles. This has been made possible by the introduction of extraction protocols using SDS with sonication, which are normally capable of extracting upwards of 95% of the total protein without addition of a reducing agent.

Based on earlier studies by Bietz (1985) and Huebner and Bietz (1985), Dachkevitch and Autran (1989) showed that 0.1M sodium phosphate buffer pH 6.9 containing 2% SDS could extract between 55 and 90% of the total flour protein from a set of 454 bread wheat lines. Extraction conditions were carried out at 60° C. This heat treatment resulted in consistent SE-HPLC profiles when run at different times after extraction. Later studies have emphasized the importance of heat treatment of extracts at about 60° C to provide consistent SE-HPLC results by presumably denaturing proteases without impacting on the elution characteristics of the major protein fractions (Larroque *et al* 2000).

Soon after this work, Singh and co-workers (1990) demonstrated that almost complete extraction of wheat flour proteins could be achieved with 0.05 M phosphate buffer pH 6.9 containing 2% SDS buffer, using sonication in a single step in the absence of reducing agents. SE-HPLC on a Waters Protein-Pak 300 column showed the presence of three peaks of decreasing size attributed to glutenin, gliadin and albumin/globulin proteins, respectively. In 1993, Guta *et al* (1993) introduced a two stage micro and macro extraction procedure that included an initial extraction in 0.05 M phosphate buffer pH 6.9 containing 0.5% SDS followed by sonication of the residue in the presence of the same buffer. Comparison of SE-HPLC profiles (Fig. 1) of these extracts to the single extraction procedure of Singh and co-workers (1990) at 214 nm, showed an enrichment of the smaller sized fractions (gliadin and albumin/globulin) and a relatively small amount of glutenin (called extractable glutenin) in the non-sonicated extract. In contrast, the second (sonicated extract) showed a much larger peak containing the largest sized (called unextractable glutenin) fraction eluting near the void volume and a single much smaller peak eluting at a size corresponding to gliadin. These extraction procedures, or modifications thereof, have been particularly useful in studying the relationships between wheat protein size distribution and inherent wheat processing quality (see reviews by Autran 1994 and Southan and MacRitchie 1999). Recently, Ueno et al (2002) introduced a two step extraction procedure using dilute acetic acid without and then with sonication that also extracts over 95% of wheat proteins. SE-HPLC and flow FFF analysis showed that the initial dilute acetic acid extraction removed primarily monomeric proteins while the second acetic acid extract with sonication removed primarily polymeric proteins.

Singh et al (1990) demonstrated that both sonication time and sonifier power strongly influences the extractability of the larger glutenin protein fraction. Sonication for 15-30 sec at a medium power setting (10-12 watts) with a Branson 20 kHz sonifier gave almost complete extraction without extensive shear degradation of the glutenin molecules.

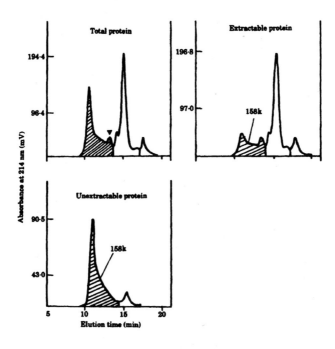

Fig. 1. SE-HPLC separations of total, extractable and unextractable flour protein on a Waters Protein Pak-300 SEC column. Reprinted from Gupta et al, ©1993, with permission from Elsevier Science.

121

Longer sonication and/or higher power setting did not increase extraction but resulted in extensive shear degradation of the glutenin polymers as shown by a decrease in the glutenin peak and an increase in gliadin and albumin/globulin peaks. More recent studies by Singh and MacRitchie (2001), and references cited within, suggest that milder sonication conditions increase extractability by reducing the molecular size of the largest glutenin polymers through shear breakage of inter-chain disulfide bonds near the midpoint of the very large polymers, similar to that which occurs during dough mixing. More intense and/or longer sonication further reduces the size of these and smaller polymers resulting in the release of much smaller polymers and some monomers. For all extraction procedures involving sonication, it is important that initial studies be carried out to monitor the impact of sonication time and intensity to optimize extraction rate while minimizing polymer degradation. For micro extraction procedures, sonication can also increase temperatures rapidly and cooling may be required to keep temperatures below 60°C to prevent denaturation (Singh *et al* 1990).

SE-HPLC analysis of wheat proteins
    SEC is the simplest form of HPLC since only a single pump and eluent is required. Excellent reviews on SE-HPLC equipment and techniques are available in several books and monographs (Autran 1994, Wu 1995, Neue 1997) as well as in this monograph (see Chapter 4). In addition to column selection, mobile phase and extraction conditions, which are discussed earlier in this chapter, a number of factors need consideration when optimizing the fractionation of wheat proteins by SE-HPLC.
    Prior to analysis, it is critical that extracts are filtered to remove any particulate matter that can clog up the column resulting in higher back pressure and shorter column life. Syringe filtering samples through a 0.45 μ glass fiber syringe filter in combination with an in-line 0.2 μ in-line filter just prior to the column removes any particulate matter and can negate the requirement for a guard column. A 1.0 μ syringe filter piggy backed on top of the 0.45 μ syringe filter can improve filtration speed by removing the larger particles and reducing clogging in the finer filter.
    The amount and concentration of protein that is injected into the SE-HPLC column can have a strong influence on resolution. High protein concentrations, particularly for larger proteins, should be avoided to reduce shear degradation and sample distortion due to "viscous fingering" while overall sample loading should represent less than 5% of the total column bed volume to prevent peak broadening. For total wheat protein extracts, about 200 μg protein in 20 μl is a good starting target (Singh *et al* 1990) for analytical size columns. It is generally recommended that less protein should be loaded for very high molecular weight proteins relative to lower molecular weight proteins (Neue 1997). Thus, less protein should be injected for extracts containing more glutenin while larger amounts of

protein can be injected for extracts containing gliadin and or albumin/globulin.

Most SE-HPLC studies of wheat proteins do not cite column temperature, which probably indicates that runs were carried out at room temperature. However, it is generally recommended that controlled temperatures above ambient be used to decrease viscosity and improve reproducibility (Neue 1997). Good resolution of wheat proteins have been obtained using temperatures of 30-40°C (Huebner *et al* 1997, Bean *et al* 1998, Lindsay *et al* 2000, Ueno *et al* 2002).

For analytical columns, flow rates for wheat protein fractionation have been varied from about 0.2 ml/min to 2 ml/min. Most work cites flow rates of 0.5 ml/min with a separation time of less than 30 min. Slower flow rates can improve resolution but reduce sample throughput. At high flow rates resolution is somewhat compromised but run time can be reduced to about 10 min (Larroque and Békés 2000).

Calibration and on-line post-column analysis of wheat proteins

Almost all SEC post-column analysis of proteins involves the use on-line UV detectors. Measurement of absorption of peptide bonds at 210-220 nm is now almost universal since these wavelengths provide high sensitivity and allow an accurate estimate of concentration.

As noted earlier, fractionation by SEC is based upon the hydrodynamic diameter of the component, not the molecular weight. However, for globular proteins, reasonable estimates of molecular weights can be obtained by calibrating columns against absorption peaks obtained for standard globular proteins of known molecular weight. Plotting of elution volume (V_e), the ratio of elution volume to the void volume (V_e/V_o) or preferably, the average partition coefficient (K_{av}) of these peaks against the logarithm of molecular weight of the standards gives a straight line for proteins that do not elute near the void or total permeation volume (Andrews 1965, Himmel *et al* 1995). K_{av} is given by

$$K_{av} = (V_e - V_o)/(V_t - V_o)$$

where V_t is the total permeation volume. More modern theoretical models that consider variation in solute and bead pore size have also been developed to calibrate columns (Himmel *et al* 1995) but are not in general use and are beyond the scope of this review. SEC has been useful for estimating the molecular weights of monomeric wheat protein fractions including gliadins, albumins and globulins as reviewed by Bietz (1986) and by Autran (1994). Although the size of polymeric glutenin fractions have also been estimated by SEC, as outlined in the reviews cited above, these results need to be viewed with caution since in most cases, elution occurs near or at the void volume where molecular weight estimates are difficult. Furthermore, studies suggest that glutenins have more extended rather than globular structures (for review see Gianibelli *et al* 2001), leading to greater retention and overestimation of their size. The lack of suitable globular standard proteins for calibration above about 700 kD also is an impediment. Protein calibration sets are available from a number of manufacturers. Some of the common proteins used as outlined by Bangur *et al* (1997) include myoglobin (17,000), chymotrypsin (42,600), ovalbumin (44,000), bovine serum albumin (67,000), γ-globulin (158,000), catalase (230,000), ferritin (440,000) and thyroglobulin (670,000).

Another approach that has become available for on-line SEC post column monitoring analysis is multiangle laser light scattering (MALLS). The combination of SEC-MALLS allows the determination of absolute weight average molecular weight (MW_w) and radius of gyration ($<R_G^2>_w^{0.5}$) of the fractionation curve for protein and/or protein complexes. The initial size separation by SEC is critical to obtaining useful information since MW_w is strongly biased towards the largest components in mixtures. In addition to the MALLS instrument, an on-line differential refractive index detector is required to monitor concentration changes (dn/dc) during measurements. The theory and use of SEC-MALLS has been given in several comprehensive reviews (Wyatt 1993, Astafieva *et al* 1996, van Dijk and Smit 2000). Lookhart (1997) reported the use on-line of MALLS to determine the molecular weight and radius of gyration of reduced glutenin

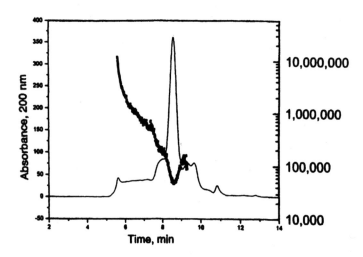

Fig. 2. Molecular weight distribution of a 1% SDS/0.05 M sodium phosphate pH 7.0 wheat protein extract eluted on a Biosep SEC 4000 column using MALLS. From Bean and Lookhart (2001) with permission.

subunits separated by RP-HPLC. Values for molecular weight agreed well with those calculated from cDNA sequences. Recently, Bean and Lookhart (2001) and Carceller and Aussenac (2001) reported the use of SEC-MALLS to characterize wheat protein extracts.

Figure 2 shows the molecular weight distribution of a 1% SDS/0.05 M sodium phosphate pH 7.0 wheat protein extract eluted on a Biosep SEC 4000 column (Bean and Lookhart 2001). The wide range of MW, particularly in the polymeric region, is clearly evident. Carceller and Aussenac (2001) used SEC-MALLS to study the size distribution of wheat polymeric glutenins extracted and purified in aqueous 50 and 70% 1-propanol containing 2% SDS with and without sonication. A PL aquagel-OH 60 column with a very high exclusion limit was used to improve resolution of the large polymeric proteins.

Figure 3 shows the SEC elution (214 nm) and MALLS MW profiles for total solubilized (by sonication) glutenin for the variety Soissons. MW_w values range from about 1×10^5 to well above 1×10^7 are evident. From these studies it is clear that SEC-MALLS shows much potential in improving our understanding of wheat protein size and shape and its influence on quality.

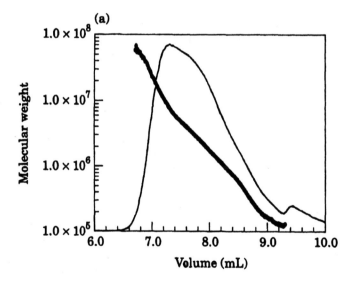

Fig. 3. Molecular weight distribution of total solubilized glutenin of Soissons eluted on PL aquagel-OH 60 using MALLS. Reprinted from Carceller and Aussenac, ©2001, with permission from Elsevier Science.

Flow Field-Flow Fractionation

Principles

Flow field-flow fractionation (flow FFF) is one of a family of techniques that allows fractionation of macromolecules and colloidal material by size related parameters. Separation is achieved by elution of components through a thin ribbon-like channel that is subjected to a field (cross-flow, centrifugal, electrical or thermal) perpendicular to the channel flow (for review see Giddings and Caldwell 1989, Schimpf et al 2000). Components showing greater response to the force exerted by the field are displaced further from the center of the channel flow and toward the channel wall. Since laminar flow through thin channels produces a parabolic flow profile, component elution order is determined by their relative displacement. Fractionation range, resolution and run time can be adjusted by varying both the channel flow and the strength of the field.

In flow FFF, components are partitioned by introducing a cross-flow of eluent perpendicular to the channel flow. In symmetrical flow FFF, the cross-flow is introduced through a porous frit embedded in the upper channel (depletion) wall and exits through a semipermeable membrane placed over a porous frit embedded in the lower (accumulation) wall (Giddings 1993). In asymmetrical flow FFF, the upper channel wall is solid, and cross-flow is generated by directing a portion of the channel flow through the accumulation wall (Litzén et al 1993). In normal mode flow FFF, where macromolecules or colloidal material are less than approximately 1 μ in diameter, displacement of components towards the accumulation wall is opposed by diffusion. Smaller components with higher diffusion coefficients reach equilibrium farther from the accumulation wall where the channel flow rate is higher and are eluted faster than larger components with lower diffusion coefficients. Particles larger than about 1 μ are separated in steric mode where elution order is opposite to normal mode. For normal mode flow-FFF, samples should be filtered through 1 μ or finer filters to eliminate particles that elute in steric mode to simplify interpretation of fractograms.

The theory of FFF and its application to the fractionation of macromolecules, colloidal and particulate materials has been reviewed by Giddings (1993). For normal flow FFF, retention time (t_r) is related to the diffusion coefficient (D) of the component, the channel thickness (w), the cross-flow rate (V_c) and the channel flow rate (V) by the equation

$$t_r = w^2 V_c/6DV$$

When this equation is combined with the Stokes-Einstein equation, $D = kT/3\pi\eta d_s$, retention time can be related to the hydrodynamic (Stokes) diameter (d_s) by

126

$$t_r = d_s \pi \eta w^2 V_c / 2kTV$$

where η is the viscosity of the carrier, k is Boltzmann's constant and T is the absolute temperature. Theoretically both the diffusion coefficient and the Stokes diameter can be determined from 1st principles without the need for calibration. In practice, it is often simpler to calibrate channels using standards with known diffusion coefficients or derived Stokes diameter values. This yields straight line relationships between t_r and d_s or 1/D that are applicable to all types of components providing the same condition are used (flow rates, carrier, etc.).

Flow FFF has been used to fractionate proteins and protein complexes and provide information on their size properties (Giddings 1993, Li and Hansen 2000). This technique is especially suited to larger proteins or complexes thereof since, unlike SEC methods, the exclusion size is effectively unlimited. The structure of macromolecules sensitive to shear can also be maintained because fractionation in FFF occurs in the absence of a solid support.

Methods and Conditions

Flow FFF channels and related equipment as well as techniques vary widely. In the following sections, descriptions of symmetrical and asymmetrical flow FFF equipment and techniques are given. An automated symmetrical flow FFF technique developed for wheat protein fractionation is also described. For a more in-depth description of FFF equipment and techniques, the reader is referred to the comprehensive book edited by Schimpf *et al* (2000).

Channels and related equipment

There are two basic types of channels used in flow FFF. Symmetrical channels consist of two plexiglass plates with embedded porous frits separated by a 100-500 μ rectangular spacer with "V" shaped ends to form a thin channel as shown in Fig. 4.

A semi-permeable membrane is placed over the lower (accumulation) wall to prevent macromolecules from exiting the channel through the frit. Eluent and samples for fractionation are introduced through an opening in the inlet end of the frit into the channel and exit through the outlet end of the frit. Cross-flow is introduced through openings in the upper plexiglass above the upper porous frit and exits through openings in the lower plexiglass plate below the lower porous frit. Low pulsation pumps (dual piston HPLC pumps or syringe pumps) are used to control the channel and the cross-flow rates. A variable back pressure regulator is used on the channel flow after passage through the UV detector (for proteins) to control the distribution of channel and cross-flow. A back pressure regulator is normally not required on the cross-flow outlet since the back pressure from

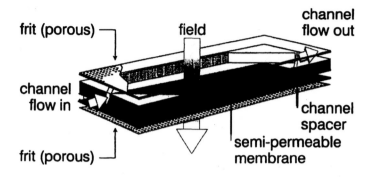

frit (porous)　　　　field　　　channel flow out

channel flow in

frit (porous)

channel spacer

semi-permeable membrane

Fig. 4. Diagram of symmetrical flow FFF channel

the membrane is higher than that of the detector. For proteins, a hydrophilic regenerated cellulose membrane with a MW cutoff of 10,000 is normally used. Samples can be introduced using a manual loop injector or an automated HPLC sampler plumbed into the channel inlet tubing.

In asymmetrical channels, the upper plexiglass plate and embedded frit is replaced by a solid glass plate. Eluent flow is pumped into the channel at the "V" shaped inlet of the spacer and is partitioned into channel flow and cross-flow by means of a variable back pressure regulator attached to the channel outlet after the UV detector. The spacer is cut in a trapezoidal shape with the narrow end towards the outlet. This design maintains the channel flow velocity despite the loss of eluent cross flow through the membrane. A second channel inlet (and pump) at the maximum width of the channel spacer is used to introduce sample.

Sample preparation and analysis by flow FFF

The same requirements apply to eluent choice for flow FFF as that for SEC discussed in section. The eluent must be a good solvent for the components and should reduce component interactions and maintain "native" conformation. Low viscosity eluents are preferable for flow FFF in order to keep back-pressure as low as possible. For most flow FFF units, back pressures of more than about 150 psi can cause leakage around the spacer. The eluent should also minimize interactions between the components and the membrane. It is normally recommended that a surfactant be used in the eluent to condition the membrane. The surfactant is believed to coat the membrane and reduce interactions with the components. FL-70 (Fisher Scientific), a mixed surfactant, is the most commonly used surfactant for flow FFF.

For wheat proteins, dilute (0.05 M) acetic acid containing 0.002% FL-70 (Stevenson and Preston 1996) and 0.25% SDS in 0.05 M sodium phosphate buffer, pH 6.8 (Wahlund *et al* 1996), have been used as eluents. With dilute acetic acid, comparison of a range of surfactants showed that FL-70 gave the best results with wheat proteins at concentrations as low as 0.0001% (Stevenson and Preston 1997). For isolation and purification of protein peaks, FL-70 can be eliminated from the eluent providing the membrane is pre-coated.

It is critical that components entering the channel undergo a relaxation/equilibrium process prior to elution. In symmetrical flow FFF, this is achieved through stop-flow. In this process, outward channel flow is 'stopped' by diverting it around the channel with a switching valve while cross-flow is maintained through the channel. Components are forced downward towards the accumulation wall membrane by the cross-flow until each component attains equilibrium through the counteracting effect of diffusion. This equilibrium is maintained when outward channel flow is resumed, and provides the basis for separation. Components with larger diffusion coefficients (smaller d_s values) equilibrate in faster flow streams of the channel parabolic flow profile and elute more quickly than larger components with lower diffusion coefficients that equilibrate in slower flow streams nearer the membrane surface. Failure to properly equilibrate samples results in poor resolution with components tending to get washed through the column with the elution front following sample injection.

Stop flow time (τ) is determined primarily by the column dimensions and the diffusion coefficient of the components. Channel thickness (w) is the most important factor since it is related to τ by its square root. Larger components take longer to equilibrate since τ is also proportional to $1/D$. In practice, it is recommended that stop flow times be adjusted such that 1-2 channel volumes of cross-flow pass across the channel and then increased (~2x) to determine if there is further improvement in resolution. The injection time required to deliver the sample to the beginning of the channel can be estimated from the tubing volume, sample volume size and channel flow rate. This time should then by varied until maximum peak height or area is attained to ensure that all of the sample has entered the channel.

In asymmetrical flow FFF, stop flow is replaced by sample focusing/relaxation since both channel and cross-flow originate from the same source. A separate inlet port through which eluent is pumped is used to inject sample into the channel at a low flow rate (normally about 0.1-0.2 ml/min) near the widest point of the trapezoid shaped channel. Through the use of switching valves, the flow direction at the channel outlet is reversed prior to injection such that the two opposing flows focus the sample downwards at a point near the accumulation wall opposite the sample input port. Focusing/relaxation normally takes about 30 sec plus sample injection time (adjusted to about 30 sec). After this step, the switching valve positions are changed back to normal flow and separation occurs.

Systematic changes should be made in these parameters to optimize resolution. Further details of this process are extensively discussed by Wahlund (2000).

Determination of optimum flow rate conditions is as much an art as a science. A fairly wide range of conditions should be tried during the optimization process. In general, higher cross-flow rates are required to resolve smaller components but lead to longer elution times and peak broadening of later eluting larger components. For wheat proteins, good resolution has been obtained with cross-flow rates 3-5 times those of the channel outlet flow rate (Stevenson and Preston 1996, Wahlund et al 1996). Outlet flows were in the 1-2 ml/min range, which provides reasonable run times well within maximum back pressure limitations. For samples with very wide size ranges, programmed field decay, where cross-flow rates are reduced with increasing elution time, can also be used to reduce peak broadening and improve resolution of the larger components (Schimpf et al 2000).

The most common detectors used for flow FFF are photometric absorbance detectors for components that have suitable chromophores such as proteins (280 or 210 nm) and refractive index detectors for those without, such as carbohydrates. More extensive use is now being made of multiangle light scattering since this technique can provide information on molecular size and hydrodynamic radius (Shortt et al 1996, Schimpf et al 2000).

In general, fractograms from flow FFF show poorer signal to noise ratios and are more difficult to obtain quantitative information from compared to other size fractionation techniques such as SE-HPLC. This is largely due to the sensitivity of flow FFF to overloading at low sample concentrations. For wheat proteins, sample loads of more than approximately 1 µg protein using a 127 µ thick symmetrical (Model F100) channel from postnova analytics USA (formally FFFractionation Inc.) caused peak broadening and loss of resolution (Stevenson and Preston 1996). Sample load is primarily dependent on the size distribution of the components (small>large) and the dimensions of the channel. Pressure pulses from pumps and, in particular, from the stop flow or focusing/relaxation step can also strongly influence detector stability at the high sensitivity setting required to obtain peaks. Insertion of pulse dampeners can help in reducing noise related to pulsation. Recirculation of cross-flow can also used to reduce noise by providing better pressure control (Li et al 1998).

A number of improvements in equipment design and techniques introduced in recent years have enhanced the resolution of flow FFF (see reviews by Li et al 1998, Schimpf et al 2000). The introduction of a frit inlet in the upper channel wall at a point just beyond the normal channel (sample) inlet allows the use of hydrodynamic relaxation. In this process, a high ratio (7-10/1) of frit inlet to channel inlet flow is used to force the incoming sample downwards to the accumulation wall. Hydrodynamic relaxation eliminates the need to divert channel flow used in stop-flow relaxation, resulting in much better pressure control and detector stability. Since all components reside near the accumulation wall during elution, a frit outlet placed at the end of the upper channel can also be used to remove 90% or more of the channel outlet flow without loss of components. The resulting large increase in component concentration of the outlet flow increases detector peak height and signal to noise ratio. When channels with both frit inlet and frit outlet are used, frit flow, in addition to cross flow, can be recirculated to further improve pressure balance.

Flow FFF of wheat proteins

Flow FFF was first applied to the size fractionation of wheat proteins by Stevenson and Preston and by Wahlund *et al* in 1996 using symmetrical or asymmetrical flow techniques, respectively. In both studies, unreduced glutenin components obtained by sonication using dilute HCl, after removal of dilute salt, 70% ethanol and dilute acetic acid extractable proteins, or by sequential extraction with increasing concentrations of HCl, showed a wide size range. Stokes diameters (d_s) of these components ranged from about 10 to well over 40 nm. Salt soluble Osborne albumins and globulins showed peaks with d_s values of 4.5, 7.2 and 10.9 nm while alcohol soluble Osborne gliadins showed a major peak at 7.4 nm and a minor peak at 9.3 nm. After removal of these fractions, the dilute acetic acid soluble "glutenin" fraction showed peaks corresponding to monomeric (non-glutenin) proteins at 7.4 and 9.9 nm and a broad polymeric glutenin peak at 16.9 nm (Stevenson and Preston 1996). Sequential extraction of flour with increasing concentrations of HCl showed a progressive increase in the size of components with later fractions (larger polymeric glutenins) showing peaks with d_s values of about 30 nm (Wahlund *et al* 1996). Both of these studies clearly demonstrated the extremely large size (probably in the 10,000,000 range) of the largest glutenin polymers and the promising potential of flow FFF for the characterization of the polymeric glutenin proteins in wheat flour. In more recent studies, flow FFF has been used by Larroque *et al* (1999) to assess the effect of sonication on the size distribution of glutenin polymers during extraction and by Beasley *et al* (2001) to study in vivo glutenin subunit oxidation.

Stevenson *et al* (1999) reported the first automated symmetrical flow FFF method and its application to the study of wheat and other proteins. In addition to automation, sensitivity was greatly increased through frit outlet (FO) sample concentration and baseline noise was reduced by replacing stop-flow relaxation with frit inlet (FI) hydrodynamic relaxation and by recirculating frit and cross-flows. This method resulted in improved resolution of wheat protein fractograms, more accurate estimates of elution times for the determination of component size, and much more accurate integration of peak areas for quantification.

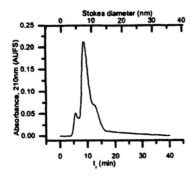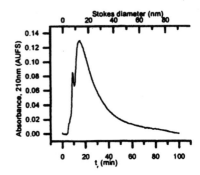

Fig. 5. Typical fractograms for AN (left) and AS (right) extracts of wheat flour. From Ueno *et al* (2002) with permission.

Figure 5 shows FIFO flow FFF fractograms of Katepwa wheat flour fractions obtained by extraction with 0.05 M acetic acid (AN) to remove primarily monomeric proteins followed by sonication of the residue with the same solvent (AS) to remove primarily polymeric proteins (Ueno *et al* 2002). Almost all of the AN fraction shows d_s values consistent with monomer (< than 10 nm) while the AS fraction shows d_s values in the polymeric size range (>10 nm) with some values well over 50 nm.

Automated FIFO FFF has been used to confirm the strong positive relationship between the proportion of larger acetic acid unextractable wheat glutenin proteins and dough strength, to assess the size characteristics of SE-HPLC fractions and to study changes in wheat proteins during mixing with and without oxidants (Preston *et al* 2001, Ueno *et al* 2002). Recent studies combining flow FFF with multiangle light scattering by S. You in the author's laboratory (unpublished data) have shown that the MW of the dilute acetic acid unextractable glutenin protein, extracted by sonication in the same buffer, range from 0.7 to about 20,000,000.

References

Ammar, K., Kronstad, W.E. and Morris, C.F. 2000. Breadmaking quality of selected durum wheat genotypes and its relationship with high molecular weight glutenin subunits, allelic variation and gluten protein polymeric composition. Cereal Chem. 77:230-236.

Andrews, P. 1965. The gel-filtration behavior of proteins related to their molecular weights over a wide range. Biochem. J. 96:595-606.

Astafieva, I.V., Eberlein, G.A., and Wang, Y.J. 1996. Absolute on-line molecular mass analysis of basic fibroblast growth factor and its multimers by reversed-phase liquid chromatography with multi-angle laser light scattering detection. J. Chromatogr. A. 740:215-229.

Aussenac, T., Carceller, J.-L. and Kleiber, D. 2001. Changes in SDS solubility of glutenin polymers during dough mixing and resting. Cereal Chem. 78:39-45.

Autran, J.-C. 1994. Size-exclusion high-performance liquid chromatography for rapid examination of size differences of cereal proteins. Pp 326-372 in High-Performance Liquid Chromatography of Cereal and Legume Proteins. (J.E. Kruger and J.A. Bietz, eds.), Am. Assoc. Cereal Chem., St. Paul, MN.

Bangur, R., Batey, I.L., McKenzie, E. and MacRitchie, F. 1997. Dependence of extensograph parameters on wheat protein composition measured by SE-HPLC. J. Cereal Sci. 25:237-241.

Batey, I.L., Gupta, R.B. and MacRitchie, F. 1991. Use of size-exclusion high-performance liquid chromatography in the study of wheat flour proteins: An improved chromatographic procedure. Cereal Chem. 68:207-209.

Bean, S.R. and Lookhart, G.L. 2001. Factors influencing the characterization of gluten proteins by size-exclusion chromatography and multiangle laser light scattering (SEC-MALLS). Cereal Chem. 78:608-618.

Bean, S.R., Lyne, R.K., Tilley, K.A., Chung, O.K. and Lookhart, G.L. 1998. A rapid method for quantitation of insoluble polymeric proteins in flour. Cereal Chem. 75:374-379.

Beasley, H L., Blanchard, C. L. and Békés, F. 2001. Preparative method for in vivo production of functional polymers from glutenin subunits of wheat. Cereal Chem. 78:464-470.

Beckwith, A.C., Nielson, H.C., Wall, J.S. and Huebner, F.R. 1966. Isolation and characterization of a high-molecular-weight protein from wheat gliadin. Cereal Chem. 43:14-28.

Bietz, J.A. 1985. High performance liquid chromatography: How proteins look in cereals. Cereal Chem. 62:210-212.

Bietz, J.A. 1986. High-performance liquid chromatography of cereal proteins. Pp 105-170 in Advances in Cereal Science and Technology, Vol. VIII (Y. Pomeranz, ed.), Am. Assoc. Cereal Chem., St. Paul, MN.

Carceller, J.-L. and Aussenac, T. 2001. Size characterization of glutenin polymers by HPSEC – MALLS. J. Cereal Sci. 33:131-142.

Ciaffi, M., Tozzi, L. and LaFiandra, D. 1996. Relationship between flour protein composition determined by size-exclusion high-performance liquid chromatography and dough rheological parameters. 1996. Cereal Chem. 73:346-351.

Cornec, M., Popineau, Y. and Lefebvre, J. 1994. Characterisation of gluten subfractions by SE-HPLC and dynamic rheological analysis in shear. J. Cereal Sci. 19:131-139.

Dachkevitch, T. and Autran, J.-C. 1989. Prediction of baking quality of bread wheats in breeding programs by size-exclusion high-performance liquid chromatography. Cereal Chem. 66:448-456.

Gianibelli, M.C., Larroque, O.R., MacRitchie, F. and Wrigley, C.W. 2001. Biochemical, genetic, and molecular characterization of wheat glutenin and its component subunits. Cereal Chem. 78: 635-646.

Giddings, J.C. 1993. Field-flow fractionation: Analysis of macromolecular, colloidal, and particulate materials. Science 260:1456-1465.

Giddings, J.C. and Caldwell, K.D. 1989. Field flow fractionation. Pp 867-938 in Physical Methods of Chemistry (B.W. Rossiter and J.F. Hamilton, eds.), John Wiley & Sons, New York, NY.

Gupta, R.B., Khan, K. and MacRitchie, F. 1993. Biochemical basis of flour properties in bread wheats. I. Effects of variation in the quantity and size distribution of polymeric protein. J. Cereal Sci. 18:23-41.

133

Hamada, A.S., McDonald, C.E. and Sibbitt, L.D. 1982. Relationship of protein fractions of spring wheat flour to baking quality. Cereal Chem. 59:296-301.

Himmel, M.E., Baker, J.O. and Mitchell, D.J. 1995. Size exclusion chromatography of proteins. Pp 409-428 in Handbook of Size Exclusion Chromatography. (C. Wu, ed.), Marcel Dekker, Inc., New York, NY.

Huebner, F.R. and Bietz, J.A. 1985. Detection of quality differences among wheats by high-performance liquid chromatography. J. Chromatogr. 327:333-342.

Huebner, F.R. and Wall, J.S. 1974. Gel-filtration and ion-exchange chromatography. Cereal Chem. 51:228-240.

Huebner, F.R. and Wall, J.S. 1976. Fractionation and quantitative differences of glutenin from wheat varieties varying in baking quality. Cereal Chem. 53:258-269.

Huebner, F.R. and Wall, J.S. 1980. Wheat glutenin: Effect of dissociating agents on molecular weight and composition as determined by gel filtration chromatography. J. Agric. Food Chem. 28:433-438.

Huebner, F.R., Nelson,T.C., Chung, O.K. and Bietz, J.A. 1997. Protein distributions among hard red winter wheat varieties as related to environment and baking quality. Cereal Chem. 74:123-128.

Larroque, O.R. and Békés, F. 2000. Rapid size-exclusion chromatography analysis of molecular size distribution for wheat endosperm protein. Cereal Chem. 77:451-453.

Larroque, O.R., Daqiq, L., Islam, N. and Békés, F. 1999. Assessing the unaltered molecular size distribution of wheat polymeric protein. pp 182-186 in Cereals 99: Proceedings of the 49th Australian Cereal Chemistry Conference, Melbourne, Australia, September 12-16, 1999. (J. F. Panozzo, M. Ratcliffe, M. Wootton and C. W. Wrigley, eds.), Cereal Chem. Division, Royal Aust. Chem. Instit., Melbourne.

Larroque, O.R., Gianibelli, M.C., Gomez Sanchez, M. and MacRitchie, F. 2000. Procedure for obtaining stable protein extracts of cereal flour and whole meal for size-exclusion HPLC analysis. Cereal Chem. 77:448-450.

Li, P., Hansen, M. and Giddings, J.C. 1998. Advances in frit-inlet and frit-outlet flow field-flow fractionation. J. Microcolumn Separations 10:7-18.

Li, P. and Hansen, M. 2000. Protein complexes and lipoproteins. Pp 433-470 in Field-Flow Fractionation Handbook. (M. E. Schimpf, K. Caldwell and J. C. Giddings, eds.), Wiley-Interscience, New York, NY.

Linarès, E., Larré, C., Lemeste, M. and Popineau, Y. 2000. Emulsifying and foaming properties of gluten hydrolysates with an increasing degree of hydrolysis: Role of soluble and insoluble fractions. Cereal Chem. 77:414-420.

Lindsay, M.P., Tamas, L., Appels, R. and Skerritt, J.H. 2000. Direct evidence that the number and location of cysteine residues affect glutenin polymer structure. J. Cereal Sci. 31:321-333.

Litzén, A., Walker, J.K., Krischollek, H. and Wahlund, K.-G. (1993) Separation and quantitation of monoclonal antibody aggregates by asymmetrical flow field-flow fractionation and comparison to gel permeation chromatography. Anal. Biochem. 212:469-480.

Lookhart, G.L. 1997. New methods helping to solve the gluten puzzle. Cereal Foods World 42:16-19.

Neue, U.D. 1997. HPLC Columns: Theory, Technology, and Practice. Wiley-VCH, New York, NY.

Nightingale, M.J., Marchylo, B.A., Clear, R.M., Dexter, J.E. and Preston, K.R. 1999. Fusarium head blight: Effect of fungal proteases on wheat storage proteins. Cereal Chem. 76:150-158.

Pasaribu, S.J., Tomlinson, J.D. and McMaster, G.J. 1992. Fractionation of wheat flour proteins by size exclusion-HPLC on an agaraose-based matrix. J. Cereal Sci. 15:121-136.

Potschka, M. 1988. Size-exclusion chromatography of polyelectrolytes. Experimental evidence for a general mechanism. J. Chromatography 441:239-260.

Prasada Rao, U.J.S. and Nigam, S.N. 1988. Chromatography of glutenin on Sepharose CL-4B in dissociating solvents: Molecular weight composition of covalently bonded glutenin. Cereal Chem. 65:373-374.

Preston, K.R., Lukow, O.M. and Morgan, B. 1992. Analysis of relationships between flour quality properties and protein fractions in a world wheat collection. Cereal Chem. 69:560-567.

Preston, K.R., Stevenson, S.G. and Takatsu, K. 2001. Effects of mixing time and oxidants on the extractability and flow FFF size distribution profiles of proteins from a Canadian hard red spring wheat flour. pp 386-390 in Cereals 2000: Proceedings of the 11th ICC Cereal and Bread Congress and of the 50th Australian Cereal Chemistry Conference, Surfers Paradise, Australia, September 11-14, 2000. (M. Wootton, I. L. Batey and C. W. Wrigley, eds.), Cereal Chem. Division, Royal Aust. Chem. Instit., Melbourne.

Schimpf, M.E., Caldwell, K. and Giddings, J.C. 2000. Field-Flow Fractionation Handbook. Wiley-Interscience, New York, NY.

Shortt, D.W., Roessner, D. and Wyatt, P.J. 1996. Absolute measurement of diameter distributions of particles using a multiangle light scattering photometer coupled with flow field-flow fractionation. Am. Lab. 28:21-28.

Singh, N.K., Donovan, G.R., Batey, I.L. and MacRitchie, F. 1990. Use of sonication and size-exclusion high-performance liquid chromatography in the study of wheat flour proteins. I. Dissolution of total proteins in the absence of reducing agents. Cereal Chem. 67:150-161.

Singh, H. and MacRitchie, F. 2001. Use of sonication to probe wheat gluten structure. Cereal Chem. 78:526-529.

Southan, M. and MacRitchie, F. 1999. Molecular weight distribution of wheat proteins. Cereal Chem. 76:827-836.

Stevenson, S.G. and Preston, K.R. 1996. Flow field-flow fractionation of wheat proteins. J. Cereal Sci. 23:121-131.

Stevenson, S.G. and Preston, K.R. 1997. Effects of surfactants on wheat protein fractionation by flow field-flow fractionation. J. Liq. Chrom. & Rel. Technol. 20:2835-2842.

Stevenson, S.G., Ueno, T. and Preston, K.R. 1999. Automated frit inlet/frit outlet flow field-flow fractionation for protein characterization with emphasis on polymeric wheat proteins. Anal. Chem. 71:8-14.

Ueno, T., Stevenson, S.G., Preston, K.R., Nightingale, M.J. and Marchylo, B.M. 2002. Simplified dilute acetic acid based extraction procedure for fractionation and analysis of wheat flour protein by size exclusion HPLC and flow field-flow fractionation. Cereal Chem. 79:155-161.

van Dijk, J.A.P.P. and Smit, J.A. 2000. Size-exclusion chromatography-multiangle laser light scattering analysis of β-lactoglobulin and bovine serum albumin in aqueous solution with added salt. J. Chromotogr. A. 867:105-1112.

Wahlund, K.-G., 2000. Asymmetrical flow field-flow fractionation. Pp 279-295 in Field-Flow Fractionation Handbook (M.E. Schimpf, K. Caldwell and J.C. Giddings, eds.), Wiley-Interscience, New York, NY.

Wahlund, K.-G., Gustavsson, M., MacRitchie, F., Nylander, T. and Wannerberger, L. 1996. Size characterisation of wheat proteins, particularly glutenin, by asymmetrical flow field-flow fractionation. J. Cereal Sci. 23:113-119.

Wu, C. 1995. Handbook of Size Exclusion Chromatography. Marcel Dekker, Inc., New York, NY.

Wyatt, P.J. 1993. Light scattering and the absolute characterization of macromolecules. Anal. Chim. Acta 272:1-40

Chapter 7

Amino Acid and Protein Sequence Analysis

Peter Koehler
Deutsche Forschungsanstalt fuer Lebensmittelchemie (German Research
Center for Food Chemistry), Lichtenbergstrasse 4, 85748 Garching, Germany

Introduction

The structure of a protein depends on the amino acid sequence (the primary structure) which determines the molecular conformation (secondary and tertiary structure). Therefore, analysis of the amino acid sequence is important for the determination of structure-function relationships of proteins. However, sequence analysis can only be conducted on a pure protein or peptide. Efficient separation of proteins or peptides is therefore necessary prior to sequence analysis. Chromatographic techniques are the methods of choice for this purpose. Additionally, amino acid analysis can be used for the detection of unusual amino acids or amino acid derivatives that cannot be determined on a protein sequencer.

Amino Acid Analysis

Amino acid analysis provides invaluable information about the composition and quantity of a peptide or protein in a sample. Knowledge of the amount of protein present is often crucial for protein chemistry applications. When only small amounts of material are present amino acid analysis can be used for protein quantification, and determination of relative amino acid composition. The amino acid composition of a peptide or protein allows the calculation of the minimum molecular mass of the sample. In combination with the amino acid composition, the molecular mass often provides sufficient information to allow the classification or identification of a sample from a database search (http://us.expasy.org/tools/aacomp/). The amino acid composition is also an invaluable tool for determining which cleavage strategy would provide the most useful fragmentation of a particular peptide or protein. Although an apparently simple process, successful amino acid analysis requires strict attention to details.

137

Peptide Separation for Sequencing

Determination of primary sequences and post-translational modifications by classical protein sequence analysis is generally applicable to proteins or peptides that have masses of less than approximately 4000. The reason for this limitation is the inability of protein sequencing instrumentation to obtain reliable information beyond 30 or 40 residues. Gluten proteins therefore need to be split into fragments of suitable size by chemical or enzymatic methods. These fragments are then purified and their amino acid compositions and sequences determined. Fragmentation generally results in complex mixtures of peptides requiring two- or more purification steps before they are sufficiently pure for sequence analysis. Usually, the initial methods selected for peptide purification have low selectivity and high sample capacity, whereas the final purification generally has the highest selectivity. A commonly employed combination would be the use of gel filtration for the first step and reversed-phase high-performance liquid chromatography (RP-HPLC) for the second.

Automated Edman Degradation

In 1950 the Swedish scientist Pehr Edman proposed a highly efficient method for the determination of the amino acid sequences of peptides or proteins. The method is comprised of the series of three reactions that when repetitively applied result in the sequential removal of amino acids from the N-terminus of the protein. The chemistry of the method has remained largely unchanged since its inception and any progress in protein sequencing by the Edman degradation has come as a result of advances in the instrumentation. Edman and Begg (1967) published a detailed description of an automated protein sequencer that was the mainstay of most protein sequencing efforts until it was supplanted by an instrument design (Hunkapiller, 1988) that has become the basis for modern-day protein sequencing instrumentation. Advances in genomics have made it possible to very rapidly determine amino acid sequences of proteins by nucleotide sequencing of the corresponding genes. Edman degradation is now primarily used for the study of protein post-translational modifications, or in cases when nucleotide sequence information will yield no information such as in the study of gluten polymer linkages. Mass spectrometry has been used for a number of years for peptide sequence determination (Spengler *et al* 1992) and is quickly increasing in importance for rapid inference of amino acid sequences of proteins for which amino acid sequence and/or nucleotide sequence information is available (Siethoff *et al* 1999).

Amino Acid Analysis

Principles

Amino acid analysis is a deceptively easy procedure involving a number of steps that must be carefully performed. During the first stage in the process the sample is hydrolysed to its constituent amino acids and in subsequent steps the amino acids are separated, detected, and quantified.

Hydrolysis

The hydrolysis of peptides and proteins can be carried out using acid hydrolysis, alkaline hydrolysis or enzymatic hydrolysis. The standard method for proteins and peptides is acid hydrolysis with 6 mol/L hydrochloric acid for 24 h at 110°C in the absence of oxygen (Hirs *et al* 1954). These conditions are a compromise between hydrolysis time and temperature so that poor recovery of some amino acids occurs. Losses for serine, threonine and methionine are 10 – 40 %, and 50 - 100 % for cysteine, tryptophan and phosphorylated amino acids. Isoleucine and valine are particularly slow to cleave. Asparagine and glutamine are completely hydrolyzed to the corresponding acids. When an accurate estimation of the more labile amino acid as well as the more difficult-to-hydrolyze amino acids is required a hydrolysis time course with points at 24, 48 and 72 hours can be carried out and recoveries extrapolated to time zero. A very useful modification is gas-phase hydrolysis (Tarr, 1986) because it can be performed rapidly using small amounts of material. The sample is hydrolyzed by vapors of hydrochloric acid allowing higher temperatures and shorter hydrolysis times (1 hour, 155°C). When conducting gas phase hydrolysis at a temperature of 155 °C it is possible to carry out the equivalent of a 24, 48, 72, hydrolysis series with time points at one, two, and three hours. If hydrolysis is carried out in a microwave oven, the time can be further reduced to several minutes (Joergensen and Thestrup, 1995; Weiss *et al* 1998).

Antioxidants such as phenol (0.1 mol/L), thioglycolic acid (0.1 - 1 %), mercaptoethanol (0.1 %), tryptamine or disodium sulfite are added to the hydrochloric acid to prevent oxidation of sensitive residues by residual oxygen. Quantitative determination of cysteine is only possible after modification. Oxidation with performic acid converts cysteine and methionine to cysteic acid and methionine sulfone (Moore, 1963). It is also possible to reduce a protein and alkylate free thiol groups with 4-vinylpyridine (Friedman *et al* 1970) or iodoacetic acid (Inglis, 1983). The corresponding derivatives after hydrolysis are pyridylethylcysteine and carboxymethylcysteine. Before undertaking any serious work with proteins it is probably wise to carefully consider the advantages and disadvantages of the different alkylating reagents that are available.

Alkaline hydrolysis for 18 - 70 h at 110°C in 4 mol/L potassium, sodium or barium hydroxide is almost exclusively used for the determination of

139

tryptophan and sometimes for phosphorylated amino acids (Hugli and Moore, 1972).

Enzymatic hydrolysis is used only for special applications, e.g. determination of glutamine and asparagine or for post-translationally modified amino acids (Maillard-reaction) that would be destroyed during acid hydrolysis (Church et al 1984). However, enzymatic hydrolysis is difficult to perform for gluten proteins because the high proportion of proline prevents cleavage at adjacent residues.

Separation, Detection, and Quantification

Amino acids from a protein hydrolysate are generally derivatized in order to enhance their detection which may be carried out either before (pre-column) or after (post-column) separation. For post-column derivatization the native amino acids are separated by ion exchange chromatography, and the derivatization reagent is added prior to detection and after separation. Chromatography is carried out on spherical ion exchange resins (polystyrol crosslinked with 10 % divinylbenzene) in citrate buffer starting at pH 2.2. A stepwise gradient with increasing ionic strength and increasing pH is applied. The classical reagent for post-column derivatization is ninhydrin (Spackman et al 1958, Smith, 1997), that forms a purple colored compound with primary amino acids (λmax = 570nm) and an orange colored compound with secondary amino acids (λmax = 440nm). Separation, derivatization and quantification are usually combined in an apparatus called an Amino Acid Analyzer. The detection limit is approximately 100 pmol. Chromatograms as shown in Fig. 1 are obtained.

In pre-column derivatization the chromatographic behavior of the amino acids is modified by using reagents that also introduce chromophors or fluorophors. Pre-column derivatization enables separation by RP-HPLC (Fig. 2) and in some cases the sensitivity is such that as little as 50 fmoles may be detected. The exact limits of detection values are difficult to predict, because contamination during hydrolysis can lead to erroneous results (also a problem in post-column techniques). Reagents used for pre-column derivatization are ortho-phthaldialdehyde (OPA; Cronin and Hare, 1977), phenylisothiocyanate (PITC; Bidlingmeyer et al 1984), 6-aminoquinolyl-N-hydroxysuccinimidyl carbamate (AQC; Cohen and Michaud, 1993) fluorenylmethoxycarbonyl (FMOC)-chloride (Miller et al 1990), and dimethylaminoazobenzene sulfonyl (DABS)-chloride (Stocchi et al 1989). The variant using PITC has gained major importance for the pre-column approach. Numerous studies conducted by the Association of Biomolecular Resource Facilities (http://www.abrf.org/) have demonstrated that comparable results can be attained with any of the above mentioned procedures. The salient point is that strict attention must be paid to details.

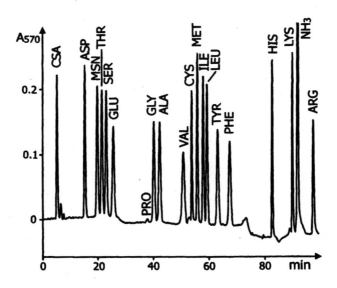

Fig. 1. Separation of an amino acid standard mixture by ion exchange chromatography and post-column derivatization with ninhydrin. CSA: cysteic acid, MSN: methionine sulfone, CYS: cystine.

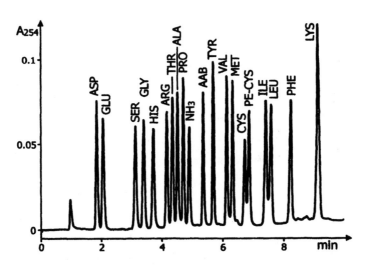

Fig. 2. RP-HPLC of PTC-amino acids after pre-column derivatization of an amino acid standard mixture with PITC. AAB: α-aminobutyric acid, CYS: cystine, PE-CYS: pyridylethyl cysteine.

141

Reagents

Hydrolysis

Hydrochloric acid 6 mol/L, freshly distilled, containing 0,1 mol/L phenol. The reagent should not be older than one week (when phenol is present).
1. Dilute conc. hydrochloric acid to 6 mol/L and distill. Collect the fraction boiling constantly at 108 °C (Alternatively, 6 mol/L constant boiling HCl is commercially available in convenient 1 ml vials).
2. Dissolve phenol in 6 mol/L hydrochloric acid to obtain a concentration of 0.1 mol/L (the phenol must be of high-quality and the solution must remain clear after its addition).

Post-Column Derivatization with Ninhydrin
1. Sodium citrate buffer, 0.1 mol/L, pH 2.2 with trifluoroacetic acid.
2. Amino acid standard mixture in citrate buffer. Concentration: 100 nmol/mL for each amino acid.

Pre-Column Derivatization with PITC
1. Conditioning solvent: ethanol/water/triethylamine, 2/2/1 (v/v/v).
2. Derivatization reagent: methanol/triethylamine/water/PITC, 7/1/1/1 (v/v/v/v).
3. Amino acid standard mixture in water. Concentration: 10 nmol/mL for each amino acid.
4. HPLC-solvents: (A) sodium acetate 0.14 mol/L/triethylamine 0.05 % (v/v)/EDTA 0.3 mg/L/pH 6.4 with glacial acetic acid, (B) 60 % (v/v) acetonitrile in water.

Apparatus

Hydrolysis
1. Glass hydrolysis vessel as outlined in Fig. 3.
2. Glass test tubes, volume 1 – 10 mL.
3. Vacuum desiccator, filled with sodium hydroxide pellets.

Post-Column Derivatization with Ninhydrin
An Amino Acid Analyzer is needed for separation, derivatization, detection and quantification.

Pre-Column Derivatization with PITC
1. HPLC system with (binary) gradient elution.
2. Column: Nova-Pak C_{18}, 3.9 × 150 mm, particle size 4 μm, pore size 60 nm.
3. UV-detector, $\lambda = 254$ nm.
4. Injection: manual or by autosampler.

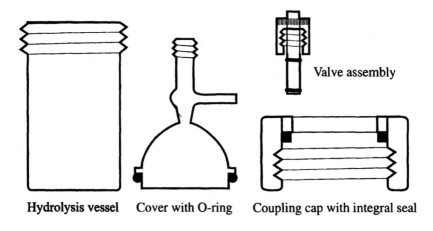

Hydrolysis vessel   Cover with O-ring   Coupling cap with integral seal

Valve assembly

Fig. 3. Hydrolysis vessel for liquid and gas-phase hydrolysis.

Procedure

Hydrolysis
1. Weigh 10 - 1000 μg of sample into a glass test tube.
2. Add 1.5 mL of phenolic hydrochloric acid to the sample.
3. Add 10 mL of phenolic hydrochloric acid into the hydrolysis vessel.
4. Put sample into the hydrolysis vessel.
5. Evacuate vessel, reventilate with nitrogen, repeat twice, evacuate again.
6. Put vessel in an oven for 24 h at 110 °C.
7. Open vessel, put sample into a desiccator containing potassium hydroxide pellets.
8. Evacuate desiccator. Wait until hydrochloric acid is evaporated.

Post-column derivatization with ninhydrin
1. Dissolve amino acids in citrate buffer. Concentration 20 - 500 nmol/mL.
2. Apply 20 μL to the amino acid analyzer.
3. Record absorbance at 570 nm and 440 nm (proline), quantify on the basis of calibration with an external standard (2 nmol/amino acid).

Pre-Column Derivatization with PITC
1. Pipet up to 25 nmol/amino acid into a glass test tube (50 × 6 mm).
2. Dry (8 - 10 Pa, 2 - 3 min), add 20 μL of conditioning solvent, dry again.
3. Add 20 μL of derivatization reagent, allow to react for 20 min at room temperature, remove reagents by evaporation.
4. Dissolve sample in HPLC starting buffer. Concentration 1 - 250 nmol/L.

5. Inject 1 - 40 μL into the HPLC system. Elute with a gradient running from 10 to 53 % B. Adapt gradient (time, slope) for optimum resolution.
6. Record absorbance at 254 nm, quantify on the basis of calibration with an external standard (250 pmol/amino acid).

Examples of Applications

Amino acid analysis has been extensively applied to wheat and the products derived from it. In most cases the variant using liquid-phase hydrolysis in 6 mol/L hydrochloric acid for 24 h at 110 °C and post-column derivatization with ninhydrin has been used. Wheats and wheat flour have been investigated by Shoup *et al* (1966). Additionally, gliadin (Bietz *et al* 1977), glutenin (Woychik *et al* 1961) and individual components of gluten fractions (Wieser *et al* 1990) have been analyzed. In these investigations amino acid analysis was used for the characterization and classification of gluten proteins and their components.

Recently, amino acid analysis has focused on the detection of specific amino acids in different materials. Köhler *et al* (1991) used amino acid analysis for the detection of cystine and cysteine in partial hydrolysates of gluten. Very recently Tilley *et al* (2001) detected post-translationally modified amino acids in dough by amino acid analysis.

Tips and Modifications

To obtain reliable quantitative data for cysteine it is important to convert this amino acid to a stable residue. Oxidation of cysteine residues with performic acid prior to hydrolysis is a convenient modification which can be carried out on a portion of the amino acid hydrolysate. Another important modification is gas-phase hydrolysis because it can be applied rapidly in cases in which very little sample is present. An excellent description of the hydrolysis equipment and of the gas phase hydrolysis procedure is to be found in Tarr (1986).

Oxidation with Performic Acid

1. To form performic acid mix 1.8 mL of formic acid (99 %), 200 μL of 30 % hydrogen peroxide, and 200 μL of methanol. Keep at 20 - 25°C for 1 h, then cool down to -20°C prior to use.
2. Cool down 10 - 1000 μg of sample in a glass test tube to -10°C.
3. Add 100 μL of performic acid and keep at -10°C for 2.5 h.
4. Add 2 mL of ice-cold water, mix, freeze, and lyophilize.
5. Perform amino acid analysis, use cysteic acid and methionine sulfone in the amino acid standard mixture.

Gas-Phase Hydrolysis

1. Pipet sample solution containing 0.1 - 5 μg of protein into a glass test tube.

2. Dry in vacuo, add 0.7 mL of constant boiling hydrochloric acid containing phenol (0.1 mol/L) to the hydrolysis vessel (not to the tubes).
3. Evacuate to 130 - 260 Pa, flush the hydrolysis vessel with argon by means of a three-way stopcock, and evacuate. Repeat this procedure twice prior to heating.
4. Heat to 155 °C for 1 h or to 110 °C for 24 h.
5. Wipe outside of tubes clean and remove residual hydrochloric acid in a desiccator containing potassium hydroxide pellets.

Peptide and Protein Separation for Sequencing

Principles

HPLC is the most commonly employed method of peptide separation prior to amino acid analysis, Edman sequencing or mass spectrometry. If proteins are subjected to these procedures, neither enzymatic nor biological activity has to be present. Therefore, HPLC can also be used for protein purification. In most HPLC chromatographic procedures peptides and proteins are separated by surface-mediated processes. The differential adsorption of peptides and proteins on the surface of the stationary phase is dependent upon the interaction between the sample, the stationary phase and the mobile phase. The various chromatographic modes differ according to the manner in which sample interaction with the stationary phase occurs. The most commonly used chromatographic modes for peptides and proteins are size-exclusion (SEC), ion-exchange (IEC), affinity (AC) and reversed-phase chromatography (RPC).

SEC discriminates between molecular species on the basis of size – differential permeation into a matrix of defined porosity. It has a relatively poor resolution power but a high sample capacity and is therefore mostly used for pre-fractionation of complex mixtures (Pharmacia, 1982; Stellwagen, 1990). IEC separates peptides and proteins on the basis of accessible surface charges and their corresponding electrostatic interaction with the sorbent. Ion exchange chromatography can be used to reduce the complexity of the samples as well as providing excellent separation of very polar peptides not suited for RPC (Rossomando, 1990). AC is based on the affinity of a peptide or protein for a specific ligand coupled to a solid support. The immobilized ligand interacts specifically with peptides and proteins that can selectively bind to it, whereas peptides and proteins that do not interact are eluted without being retained (Ostrove, 1990). However, the method of choice for peptide separation is reversed-phase high-performance liquid chromatography (RP-HPLC). This technique can also be used for protein purification. Separation is based on hydrophobic interaction of the sample with the nonpolar stationary phase in a polar, aqueous solvent. Elution is carried out by means of an increasing proportion of an nonpolar, organic solvent (modifier) displacing the adsorbed material according to its hydrophobicity (Serwe *et al* 1999; Chicz and Regnier, 1990).

RP-HPLC has the advantage that samples may be dissolved in a wide variety of agents including 6 - 8 mol/L urea or guanidinium hydrochloride making it a convenient procedure to purify samples after dissolution or after reduction and alkylation. Loop volumes as large as 5 ml can be employed with no deleterious effect on the chromatography (Vensel *et al* 1989). Perhaps the most important thing to be born in mind is that samples should not contain any agent (acetonitrile, alcohol or high concentrations of formic acid) that would cause the sample to elute in the column breakthrough. Reversed phase-HPLC is easily used with volatile solvents making it the method of choice for the preparation of samples destined for amino acid analysis, mass spectrometry and protein sequencing.

Reagents

Solvents

The excellent resolving power of RP-HPLC largely results from the use of volatile perfluorinated carboxylic acids such as trifluoroacetic acid (TFA) as ion-pairing agents in the solvent (Bennett *et al*, 1980). Peptides and proteins are charged molecules at most pH values and the presence of counterions influences their chromatographic behavior. At low pH (~2.0) the carboxyl groups of the peptides and proteins are protonated and ion pairs are formed between TFA and positively charged peptides and proteins thereby increasing their interaction with the stationary phase. Acetonitrile is the most commonly used organic modifier because of its low viscosity (low back pressure), high UV-transparency at low wavelengths and high volatility. The following solvent system can be used for most purposes:
1. Solvent A: 0.1 % (v/v) TFA in water.
    Solvent B: 0.08 % (v/v) TFA in acetonitrile. A slight decrease in TFA concentration will prevent an excessive rise in the baseline while having minimal effect on chromatography.
    If gradient systems requiring a higher pH are needed to take advantage of differences in charges on the peptides, the following buffered solutions are convenient (Köhler *et al* 1991; Serwe *et al* 1999):
2. Solvent A: 0.02 mol/L ammonium acetate (NH_4Ac), pH 5 – 6.
    Solvent B: 0.02 mol/L NH_4Ac in 50 % (v/v) aqueous acetonitrile.
3. Solvent A: 0.01 mol/L triethylammonium formate (TEAF) pH 3.5 – 5.
    Solvent B: 0.008 mol/L TEAF in 40 % (v/v) aqueous acetonitrile.

Rechromatography of peptide or protein subfractions can be carried out either by altering the pH of the solvent, replacing acetonitrile with methanol or isopropanol or by the use of reversed phase columns with different selectivity. Experiments should be planned so that the final separation is carried out in a volatile solvent system. The sample solvent must be

146

compatible with the mobile phase solvent so that the sample or buffer does not precipitate in the column packing. Due to their larger sphere of hydration potassium salts are less likely to precipitate than sodium salts, however, salt precipitation is not usually a problem as long as the salt solution is injected when the column is equilibrated with low (5 – 10 %) organic solvent. Samples should not contain more organic solvent than the initial column conditions otherwise the analyte will be eluted during sample loading. If necessary the sample can simply be diluted with 0.1 % TFA. It is usually prudent to collect the column breakthrough peak in a large tube and save it until it is known that the expected peptides or proteins are recovered.

Apparatus
1. HPLC system with binary gradient mixing.
2. Column: 10 mm × 250 mm, 4.6 mm × 250 mm or 2 mm × 150 mm, packing C_{18} silica gel, particle size 3 - 5 μm, pore size 10 – 30 nm. A guard column between the injector and the main HPLC column is strongly recommended to prevent blockage of the inlet frit or the packing bed of the column. It should be filled with the same material as the main column.
3. Column oven to control the temperature between 25 and 60 °C.
4. Hand injector or autosampler.
5. Tubings: all tubing through which the sample passes should be of 0.1 mm i.d. and as short in length as possible. Some investigators advocate the use of stainless steel tubing and fittings and stainless steel tends to be more rugged than the polyetheretherketone (PEEK) material. The use of biocompatible PEEK tubing and fittings is probably only required when working with enzymes or other proteins known to lose their activity on metal surfaces.
6. Programmable UV-Detector, cell volume 3 – 10 μL. A wavelength of 214 nm is commonly used as it closely follows the peptide bond absorption while providing minimal TFA-related baseline rise. If dual wavelength detection is possible, aromatic amino acids can be discriminated by using 254 or 280 nm as the second wavelength. It is usually prudent to first obtain a so-called "peptide-map" by doing a preliminary experiment in which 5 to 10 % of the sample is injected. Referring to the peptide map will then make it possible to anticipate the emergence of each peak during the preparative run.

Procedure

Sample Preparation
1. Use peptide solution from partial enzymatic or chemical hydrolysis or dissolve lyophilised material in an appropriate solvent, e.g. 0.1 % (v/v) TFA. Concentration: 0.1 – 5 mg/mL.

2. All the material to be injected should be dissolved and free from particles. Centrifugation (19000 g, 20 °C, 5 min) or filtration (0.45 μm) is highly recommended prior to injection.

Chromatography
1. Equilibrate HPLC system for 15 - 20 min at starting conditions.
2. Inject 5 - 500 μL of sample (sample loops of up to 5 ml can be easily used) manually or by autosampler. The maximum load should not exceed about 10 mg for 9 mm columns, 2 mg for 4.6 mm columns and 0.5 mg for 2 mm columns.
3. Conditions for 9 mm columns: TFA system: flow rate 1.5 mL/min, gradient 0 - 85 min 5 - 50 % solvent B, 85 - 100 min 50 - 95 % solvent B. NH$_4$Ac and TEAF systems: 0 - 90 min 5 - 100 % solvent B (flow rates as high as 5 ml/min can be used with these columns, but they operate quite well at lower flow rates and solvent consumption is considerably reduced).
4. Conditions for 4.6 mm columns: flow rate 0.8 - 1.0 ml/min, gradients as described above.
5. Conditions for 2 mm columns: flow rate 0.2 - 0.25 ml/min, gradients as described above.
6. Detect peptides or proteins at 214 nm (range 210 - 220 nm).
7. Fractionation: optimal pooling can only be assured by collecting the peaks by hand into small glass or plastic tubes. Automatic fraction collectors can also be used, but reproducible pooling is only possible at flow rates above 0.5 mL/min. Store peptides and proteins at -20 °C prior to rechromatography or sequence analysis.

Examples of Applications
RP-HPLC is a very important tool in gluten protein and peptide research. Other techniques such as IEC, SEC or AC have also been used, but mostly in combination with HPLC. One major application is the determination of disulfide bonds in gliadin and glutenin. Many papers related to this topic have been published (Köhler *et al* 1991, 1993; Tao *et al* 1992; Keck *et al* 1995; Müller and Wieser, 1995, 1997, Egorov *et al* 1999). The strategy was to partially hydrolyze the protein, detect and isolate cystine-containing peptides by RP-HPLC or a combination of SEC and RP-HPLC, and carry out N-terminal sequence analysis. Other applications are the localization of cysteine residues in cereal proteins (Egorov *et al* 1975; Masci *et al* 1998, 1999; Köhler and Wieser, 2000). Selective enrichment of cysteine peptides was performed by covalent chromatography, relevant fractions were then separated by RP-HPLC. Other applications encompass the identification and quantitative determination of wheat-specific peptides in chymotryptic gliadin digests for quantification of the amount of wheat in cereal products (Naegele, 1991) and the characterization of gluten proteins or fractions by peptide mapping (Wieser *et al* 1985, Du Cros, 1991).

Tips and Modifications

As complex peptide mixtures are usually present after partial hydrolysis of gluten proteins, direct identification or preparation of relevant peptides is often difficult. Therefore, pre-fractionation of the mixtures may be required. SEC prior to RP-HPLC discriminates peptides on the basis of molecular size, and is usually sufficient to decrease the complexity of mixtures. The resulting fractions are separated according to their size and can usually be resolved by RP-HPLC. For selective enrichment of peptide classes, such as cysteine-containing peptides, affinity chromatography can be used as initial step. A newly emerging methodology involves conjugation of cysteine residues with isotope coded affinity tags (ICAT), selective isolation by biotin affinity chromatography and identification by tandem mass spectrometry (Gygi *et al* 1999).

Pre-fractionation by size-exclusion chromatography
1. Dissolve peptide mixture in 0.1 mol/L acetic acid and centrifuge at 19000 g, 20 °C for 10 min. Concentration: 50 mg/mL.
2. Apply to an SEC column (25 × 900 mm) filled with Sephadex G25 fine. Maximum sample load is 1 g.
3. Elute with 0.1 mol/L acetic acid at a flow rate of 0.4 mL/min.
4. Detect peptides at 220 nm, collect manually or by fraction collector.

Automated Edman Degradation

Principles

The Edman Degradation

Edman degradation cleaves the N-terminal amino acid from a peptide or protein and prepares the modified amino acid for identification. By repeating this chemical reaction, the amino acid sequence of a peptide or protein can be determined (Edman, 1950). Three individual steps are involved: coupling, cleavage and conversion. The resulting product from this series of reactions, the phenylthiohydantoin or PTH amino acid, is subsequently identified by RP-HPLC. The reaction sequence of the Edman degradation is shown in Fig. 4.

During the alkaline coupling step phenylisothiocyanate (PITC) reacts with the free N-terminal amino group of the peptide or protein to generate a phenylthiocarbamyl (PTC) derivative. A side reaction is hydrolysis of PITC giving aniline, which reacts with PITC, yielding diphenylthiourea (DPTU). In the second step the PTC derivative is treated with liquid or gaseous TFA, to cleave the PTC-N-terminal amino acid from the rest of the peptide or protein, giving the anilinothiazolinone (ATZ) amino acid. The ATZ amino acid is extracted from the residual peptide or protein by a non-polar solvent, e.g. n-butylchloride and ethylacetate. In the final reaction step, the unstable ATZ amino acid is converted to the more stable

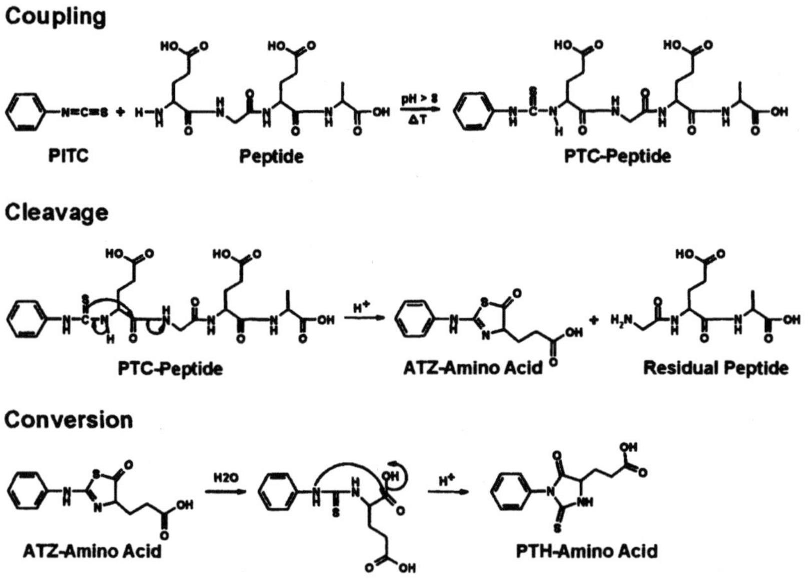

Coupling

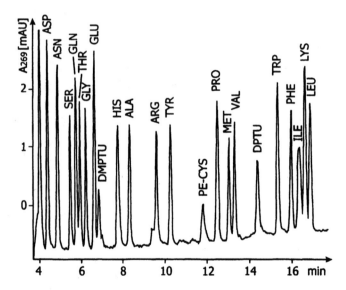

PITC + **Peptide** → **PTC-Peptide**

Cleavage

PTC-Peptide → **ATZ-Amino Acid** + **Residual Peptide**

Conversion

ATZ-Amino Acid → **PTH-Amino Acid**

Fig. 4. Reaction sequence of the Edman degradation.

Fig. 5. Separation of PTH amino acids by RP-HPLC. DMPTU: dimethylphenylthiourea, PE-CYS: pyridylethyl cysteine, DPTU: diphenylthiourea.

phenylthiohydantoin (PTH) amino acid in the presence of aqueous TFA. The shortened peptide or protein is now available for another reaction cycle.

Identification of PTH Amino Acids

The PTH amino acids produced by the Edman Degradation are then separated by RP-HPLC. Identification and quantification are carried out on the basis of the retention times and the UV-absorbance of a PTH amino acid standard mixture as shown in Fig. 5. PTH amino acids have characteristic UV-spectra with a maximum absorbance at 269 nm (Edman and Sjoquist, 1956).

Sensitivity and Efficiency of Sequence Analysis

The lower detection limit of the latest generation of automated sequencers is approximately 0.2 to 0.5 pmol of a peptide or protein but for the best results approximately 5 to 10 pmol of material should be submitted. The efficiency of the sequencing reaction is defined by the amount of PTH amino acids recovered at each cycle ("repetitive yield"). The higher this yield, the longer the amino acid sequence that can be determined. Modern sequencers have repetitive yields of ~95 %, giving sequencing results of 30 - 40 amino acid residues. The "initial yield" is the relation of the amount of PTH amino acid detected in the first cycle and the amount of protein applied for sequencing. Typical initial yields are in the range of only 50 %.

Problems

Some amino acid residues are partly or completely destroyed during the Edman procedure. Elimination of water occurs with serine (~80 %) and threonine (~50 %), tryptophan is destroyed to a great extent and cysteine (unless derivatized beforehand) is largely destroyed. The ATZ amino acids of histidine and arginine are poorly extracted because of their polarity. Therefore, sequence interpretation is often difficult because of small or missing PTH peaks or artifact peaks in the HPLC chromatograms. Modified amino acids need to be included in the PTH standard mixture and their position in the PTH chromatogram should not overlap with the other PTH amino acids. Additional factors that complicate the interpretation of protein sequence information include the fact that the Edman degradation has a cycle-to-cycle efficiency of about 93 to 95 % that, in combination with background generation, can limit the length of the protein sequencing signal that can be read. The sequencing of the gluten proteins is further complicated by the fact that they contain large proportions of proline and glutamine. Proline can be particularly troubling because it is slow to cleave and can make it extremely difficult to interpret sequencing results (Vensel and Kasarda, 1991). The large amount of glutamine present in the gluten proteins leads to rapid buildup of PTH-glutamine, further complicating sequence interpretation. A detailed discussion of some of these problems is provided in Müller et al (1998).

Reagents

Solvents and reagents for sequence analysis have to be of very high purity, otherwise the repetitive yield decreases and side reactions give artifact peaks in the HPLC chromatogram. Therefore, all reagents and solvents have to be of "sequencing grade" quality.

Apparatus

Instrumentation

Protein or peptide sequence analysis is currently carried out almost exclusively by automated protein sequencing instrumentation. Such a device consists of a solvent and reagent delivery system operated by argon pressure and controlled by electronically operated valve blocks, a reaction cartridge for the coupling and cleavage reaction, a conversion flask for PTH amino acid generation, a RP-HPLC system for on-line separation of PTH amino acids and a computer-based controller capable of providing control of the mechanical components as well as data analysis.

Modes of Operation

Two types of operation mode are commonly employed, gas phase and pulsed liquid. In the gas phase mode the coupling base and the cleavage acid are delivered as gases. The pulsed liquid mode differs from the gas phase mode only in the manner in which the cleavage acid is delivered to the reaction cartridge. A small liquid pulse of reagent, just enough to wet the sample, is delivered. The reaction cycle time is approximately 30 minutes when the sequencer is operated in the liquid phase mode and slightly longer in the gas phase mode. Regardless of the operation mode, retention of the sample in the reaction cartridge is facilitated by the use of a polymer matrix (polybrene).

HPLC System

The HPLC system for the separation of PTH amino acids has to be of highest performance. Twenty PTH amino acids and two by-products need to be separated in less than 20 min (Fig. 5). In order to achieve the increased sensitivity seen in modern-day sequencing instruments either narrow bore (2 mm i.D.) or capillary HPLC columns are used. Syringe pumps are used to reduce pulsation in the solvent that is delivered to the column and to also provide accurate gradient formation.

Sample requirements

The sample is applied onto a chemically inert glass fiber filter that is treated with polybrene so as to retain the sample in the reaction cartridge (Strickler *et al* 1984). Contamination of the sample with salts, components containing amino groups (e.g. free amino acids, Tris) or detergents interfere with the Edman degradation. Contaminants that react with PITC are usually extracted during the first cycles by the nonpolar solvents and produce artifact signals in the HPLC chromatograms. However, in these cases, sequence interpretation is still possible after two or three cycles. Polar contaminants are poorly soluble during extraction and disturb sequencing for many cycles.

Procedure
1. Coat the glass fiber filter with polybrene (as specified by the instrument manufacturer) and run sequencer for three cycles to remove contaminations.
2. Dissolve sample in a volatile solvent e.g. 0.1 % TFA. Concentration: ~50 pmol/10 μL. The amount of sample applied to the glass fiber filter will depend upon the particular instrument being used. In general, for the older instruments (ABI 477) 50 pmol would be a reasonable amount to load, however for newer instruments such as the ABI Procise 490 series below the 5 to 10 pmol should be sufficient. For dilute samples, or samples containing salts it is possible to apply them using a ProSorb® cartridge or to concentrate and desalt them with RP-HPLC (Fig. 6).
3. Run sequencer.

Examples of Applications
Automated Edman degradation has extensively been used in gluten protein analysis, especially in studies on post-translational modifications or for the characterization of proteins with a lack of sequence information. Most of the applications described in the "peptide separation" section yielded purified peptides that were characterized by the determination of their N-terminal amino acid sequences. Furthermore, N-terminal sequences of gliadins and low molecular weight subunits of glutenin have been determined by several authors (Bietz *et al* 1977; Kasarda *et al* 1983; Masci *et al* 1993, 1995). Recently, Masci *et al* (2002) have used N-terminal mixture sequencing to determine glutenin polymer composition.

Tips and Modifications

Sample Preparation
Proteins are often purified by gel electrophoresis. The N-terminal sequence can be obtained after blotting onto a polyvinylidendifluoride (PVDF) membrane which can be directly introduced into the reaction cartridge of the sequencer. Standard protocols for electroblotting can be used (Matsudaira, 1987). Because most of the transfer buffers contain amino acids these will show up as background contamination doing early sequencing cycles even after carefully washing the membrane after transfer. Salts, denaturing agents or detergents that are present in peptide or protein solutions can be removed by means of dot blotting devices as shown in Fig. 6. The device is first activated with a small amount of methanol and the

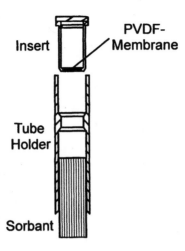

Fig. 6. *ProSorb® dot blotting device (the above copyrighted image is provided courtesy of Applied Biosystems (an Applera Corporation Business)).

peptide or protein sample applied directly onto the membrane. The peptide or protein is bound to the PVDF membrane, and excess liquid is absorbed by the cartridge. These devices can be used with samples containing high salt concentrations provided the membrane is washed sufficiently.

Blocked N-Termini

If the N-terminus of a protein or peptide is blocked, it has to be cleaved chemically or enzymatically to generate peptides with free N-termini that can be sequenced. Experimental outlines for cleavage are given by Matsudaira (1990). N-terminal blockage by pyroglutamic acid can be removed by pyrogluamate aminopeptidase (Doolittle, 1972; Podell and Abraham, 1978).

---

*ProSorb is a registered trademark of Applera Corporation or its subsidiaries in the US and certain other countries

References

Bennett, H.P.J., Browne, C.A. and Solomon, S. 1980. The use of perfluorinated carboxylic acids in the reversed-phase HPLC of peptides. J. Liq. Chromatogr. 3:1353-1365.

Bidlingmeyer, B.A., Cohen, S.A. and Tarvin, T.L. 1984. Rapid analysis of amino acids using pre-column derivatization. J. Chromatogr. 336:93-104.

Bietz, J.A., Huebner, F.R., Sanderson, J.E. and Wall, J.S. 1977. Wheat gliadin homology revealed through N-terminal amino acid sequence analysis. Cereal Chem. 54:1070-1083.

Chicz, R.M. and Regnier, F.E. 1990. High performance liquid chromatography: effective protein purification by various chromatographic modes. Meth. Enzymol. 182: 392-421.

Church, F.C., Swaisgood, H.E. and Catignani, G.L. 1984. Compositional analysis of proteins following hydrolysis by immobilized proteases. J. Appl. Biochem. 6:205-211.

Cohen, S.A. and Michaud, D.P. 1993. Synthesis of a fluorescent derivatizing reagent, 6-aminoquinolyl-N-hydroxysuccinimidyl carbamate, and its application for the analysis of hydrolysate amino acids via high-performance liquid chromatography. Anal. Biochem. 211:279-287.

Doolittle, R.F. 1972. Terminal pyrrolidonecarboxylic acid. Cleavage with enzymes. Meth. Enzymol. 25:231-244.

Du Cros, D.L. 1991. Isolation and characterization of two gamma gliadin proteins from durum wheat. J. Cereal Sci. 13:237-253.

Cronin, J.R. and Hare, P.E. 1977. Chromatographic analysis of amino acids and primary amines with o-phthalaldehyde detection. Anal. Biochem. 81:151-156.

Edman, P. 1950. Method for the determination of the amino acid sequence in peptides. Acta Chem. Scand. 4:283-290.

Edman, P. and Sjoquist, J 1956. Identification and semi-quantitative determination of phenylthiohydantoins. Acta Chem. Scand. 10:1507 - 1509.

Edman, P. and Begg, G. 1967. A protein sequenator. European J. Biochem. 1: 80-91.

Egorov, T.A., Svenson, A., Ryden, L. and Carlsson, J. 1975. A rapid and specific method for isolation of thiol-containing peptides from large proteins by thiol-disulfide exchange on a solid support. Proc. Natl. Acad. Sci. USA 72:3029-3033.

Egorov, T.A., Odintsova, T.I. and Musolyamov, A.K. 1999. Determination of disulfide bonds in gamma-46 gliadin. Biochemistry 64:294-97.

Friedman, M., Krull, L.H. and Cavins, J.F. 1970. The chromatographic determination of cystine and cysteine residues in proteins as s-β-(4-pyridylethyl)cysteine. J. Biol. Chem. 245:3868-3871.

Gygi, S.P., Rist, B., Gerber, S.A., Turecek, F., Gelb, M.H. and Aebersold, R. 1999. Quantitative analysis of complex protein mixtures using isotope-coded affinity tags. Nat. Biotechnol. 17:994-999.

Hirs, C.H.W., Stein, W.H. and Moore, S. 1954. The amino acid composition of ribonuclease. J. Biol. Chem. 211:941-950.

Hugli, T.E. and Moore, S. 1972. Determination of the tryptophan content of proteins by ion exchange chromatography of alkaline hydrolysates. J. Biol. Chem. 247:2828-2834.

Hunkapiller, M.W. 1988. Automated protein sequencing. Pages 256 - 270 in: Methods Protein Analysis. Cherry, J.P. and R.A. Barford, eds. AOCS: Champaign, IL, USA.

Inglis, A.S. 1983. Single hydrolysis method for all amino acids, including cysteine and tryptophan. Meth. Enzymol. 91:26-36.

Joergensen, L and Thestrup, H.N. 1995. Determination of amino acids in biomass and protein samples by microwave hydrolysis and ion-exchange chromatography. J. Chromatogr. A 706:421-428.

Kasarda, D.D., Autran, J.C., Lew, E.J.L., Nimmo, C.C. and Shewry, P.R. 1983. N-Terminal amino acid sequences of ω-gliadins and ω-secalins. Implications for the evolution of prolamin genes. Biochim. Biophys. Acta 747:138-150.

Keck, B., Köhler, P. and Wieser, H. 1995. Disulphide bonds in wheat gluten: cystine peptides derived from gluten proteins following peptic and thermolytic digestion. Z. Lebensm. Unters. Forsch. 200:432-439.

Köhler, P., Belitz, H.-D. and Wieser, H. 1991. Disulphide bonds in wheat gluten: isolation of a cystine peptide from Glutenin. Z. Lebensm. Unters. Forsch. 192:234-239.

Köhler, P., Belitz, H.-D. and Wieser, H. 1993. Disulphide bonds in wheat gluten: further cystine peptides from high molecular weight (HMW) and low molecular weight (LMW) subunits of glutenin and from γ-gliadins. Z. Lebensm. Unters. Forsch. 196:339-247.

Köhler, P. and Wieser, H. 2000. Comparative studies of high M_r subunits of rye and wheat. iii. localisation of cysteine residues in high M_r fractions of rye. J. Cereal. Sci. 32:189-197.

Masci, S., Lafiandra, D., Porceddu, E., Lew, E.J.L., Tao, H.P. and Kasarda, D.D. 1993. D-glutenin subunits: N-terminal sequences and evidence for the presence of cysteine. Cereal Chem. 70:581-585.

Masci, S., Lew, E.J.L., Lafiandra, D., Porceddu, E. and Kasarda, D.D. 1995. Characterization of low molecular weight glutenin subunits in durum wheat by reversed-phase high-performance liquid chromatography and N-terminal sequencing. Cereal Chem. 72:100-104.

Masci, S., D'Ovidio, R., Lafiandra, D. and Kasarda, D.D. 1998. Characterization of a low-molecular-weight glutenin subunit gene from bread wheat and the corresponding protein that represents a major subunit of the glutenin polymer. Plant Physiol. 118:1147-1158.

Masci, S., Egorov, T.A., Ronchi, C., Kuzmicky, D.D., Kasarda, D.D. and Lafiandra, D. 1999. Evidence for the presence of only one cysteine residue in the d-type low molecular weight subunits of wheat glutenin. J. Cereal Sci. 29:17-25.

Masci, S, Rovelli, L., Kasarda, D.D., Vensel, W.H., Lafiandra, D. 2002. Characterization and chromosomal localization of the C-type and low-molecular weight subunits in the bread wheat cultivar Chinese Spring. Theor. Appl. Genet. 104:422-428.

Matsudaira, P. 1987. Sequence from picomole quantities of proteins electroblotted onto polyvinylidene difluoride membranes. J. Biol. Chem. 262:10035-10038.

Matsudaira, P. 1990. Limited N-terminal sequence analysis. Meth. Enzymol. 182:602-613.

Meltzer N.M., Tous, G.I., Gruber, S. and Stein, S. 1987. Gas-phase hydrolysis of proteins and peptides. Anal. Biochem. 160:356-361.

Miller, E.J., Narkates, A.J. and Niemann, M.A. 1990. Amino acid analysis of collagen hydrolysates by reverse-phase high-performance liquid chromatography of 9-fluorenylmethyl chloroformate derivatives. Anal. Biochem. 190:92-97.

Moore, S. 1963. On the determination of cystine as cysteic acid. J. Biol. Chem. 238:235-237.

Müller, S. and Wieser, H. 1995. The location of disulphide bonds in α-type gliadins. J. Cereal Sci. 22:21-27.

Müller, S. and Wieser, H. 1997. The location of disulphide bonds in monomeric γ-type gliadins. J. Cereal Sci. 26:169-176.

Müller, S., Vensel W. H., Kasarda, D. D., Köhler, P. and Wieser, H. 1998. Disulfide bonds of adjacent cysteine residues in LMW subunits of glutenin. Journal of Cereal Science 27:109-116.

Naegele, R., Belitz, H.-D. and Wieser, H. 1991. Analysis of food and feed via partial sequences of characteristic protein components. Part 1. Isolation and identification of wheat-specific peptides from chymotryptic hydrolyzates of gliadin. Z. Lebensm. Unters. Forsch. 192:415-421.

Ostrove, S. 1990. Affinity Chromatography: General Methods. Meth. Enzymol. 182:357-371.

Pharmacia Fine Chemicals 1982. Gel filtration – theory and practice. Pharmacia Fine Chemicals AB, Uppsala, Sweden.

Podell, D. N. and Abraham, G.N. 1978. A technique for the removal of pyroglutamic acid from the amino terminus of proteins using calf liver pyroglutamate amino peptidase. Biochim. Biophys. Res. Commun. 81:176-185.

Rossomando, E.F. 1990. Ion-Exchange Chromatography. Meth. Enzymol. 182:309-317.

Serwe, M., Blüggel, M. and Meyer, H.E. 1999. High Performance Liquid Chromatography. Pages 67 - 85 in: Microcharacterization of Proteins. Kellner, R., Lottspeich, F. and H.E. Meyer, eds. Second Edition, Wiley – VCH: Weinheim, Germany.

Shoup, F.K., Pomeranz, Y. and Deyoe, C.W. 1966. Amino acid composition of wheat varieties varying widely in bread-making potentialities. J. Food. Sci. 31:94-101.

Siethoff, S., Lohaus, C. and Meyer H.E. 1999. Sequence analysis of proteins and peptides by mass spectrometry. Pages 245 - 273 in: Microcharacterization of Proteins. Kellner, R., Lottspeich, F. and H.E. Meyer, eds. Second Edition, Wiley – VCH: Weinheim, Germany.

Smith, A.J. 1997. Postcolumn amino acid analysis. Methods Mol. Biol. 64:139-146.

Spackman, D.H., Moore, S. and Stein, W.H. 1958. Automatic recording apparatus for use in the chromatography of amino acids. Anal. Chem. 30:1190-1205.

Spengler, B., Kirsch, D., Kaufmann, R. and Jaeger, E. 1992. Peptide sequencing by matrix-assisted laser-desorption mass spectrometry. Rapid Commun. Mass Spectrom. 6:105-108.

Stellwagen, E. 1990. Gel Filtration. Meth. Enzymol. 182:317-328.

Stocchi, V., Piccoli, G., Magnani, M., Palma, F., Biagiarelli B. and Cucchiarini, L. 1989. Reversed-phase high-performance liquid chromatography separation of dimethylaminoazobenzene sulfonyl- and dimethylaminoazobenzene thiohydantoin-amino acid derivatives for amino acid analysis and microsequencing studies at the picomole level. Anal. Biochem. 178:107-117.

Strickler, J.E., Hunkapiller, M.W., Wilson, K.J. 1984. Utility of the gas-phase sequencer for both liquid- and solid-phase degradation of proteins and peptides at low picomole levels. Anal. Biochem. 140:553-566.

Tao, H.P., Adalsteins, A.E. and Kasarda, D.D. 1992. Intermolecular disulfide bonds link specific high-molecular weight glutenin subunits in wheat endosperm. Biochim. Biophys. Acta 1159:13-21.

Tarr, G. E. 1986. Manual Edman sequencing system. Pages 155-194 in: Methods of protein microcharacterization: A Practical Handbook. J. E. Shively, ed. Humana Press: Clifton, NJ, USA.

Tilley, K.A., Benjamin, R.E., Bagorogoza, K.E., Okot-Kotber, B.M., Prakash, O. and Kwen, H. 2001. Tyrosine cross-links: molecular basis of gluten structure and function. J. Agric. Food Chem. 49:2627-2632.

Vensel, W.H., Lafiandra, D. and Kasarda, D.D. 1989. The effect of an organic eluent modifier and pH on the separation of wheat-storage proteins: application to the purification of γ-gliadins of Triticum monococcum L. Chromatographia 28:133-138.

Vensel, W.H. and Kasarda, D.D. 1991. Effect of cleavage conditions on Edman degradation of proline-rich proteins: application to wheat storage proteins. Pages 181-190 in: Techniques of Protein Chemistry II. J. Villifranca, ed. Academic Press: New York, NY, USA.

Weiss, M., Manneberg, M., Juranville, J.F., Lahm, H. W. and Fountoulakis, M. 1998. Effect of the hydrolysis method on the determination of the amino acid composition of proteins. J. Chromatogr. A 795:263-275.

Wieser, H., Stempfl, C. and Belitz, H-D. 1985. Peptide patterns of the glutelin fractions of different wheat varieties. Z. Lebensm. Unters. Forsch. 181:1-3.

Wieser, H. Seilmeier, W. and Belitz, H.-D. 1990. Characterization of high molecular weight subunits of glutenin separated by reversed-phase high-performance liquid chromatography. J. Cereal Sci. 12:223-227.

Woychik, J.H., Boundy, J.A. and Dimler, R.J. 1961. Amino acid composition of proteins of wheat gluten. J. Agric. Food Chem. 9:307-310.

Chapter 8

Spectroscopic Analysis of Gluten Protein Structure: Circular Dichroism and Infra-Red.

Arthur S Tatham
Long Ashton Research Station, Department of Agricultural Sciences,
University of Bristol, Long Ashton, Bristol BS41 9AF, UK

Nikolaus Wellner
Institute of Food Research, Norwich Research Park, Colney, Norwich
NR4 7UA, UK

Introduction

Circular dichroism (CD) and infra-red (IR) spectroscopies are types of absorption spectroscopy that can provide information on the conformations of proteins and other biological macromolecules.

CD is observed in chiral or asymmetric molecules where there is a difference in absorption between left and right-handed circularly polarized light. Two types of asymmetry exist in polypeptides. Firstly, all amino acids, except glycine, are asymmetric around their C_α atoms and the far-UV region (below 250nm) is therefore dominated by absorption of the amide bond. Furthermore different secondary structures (helix, sheet, turns and coil) give rise to different CD spectra. As the spectrum of a protein is directly related to its secondary structure content, the spectrum is a linear combination of the spectra of these structures; it is therefore possible to deconvolute the spectra to obtain secondary structure information. Secondly, the aromatic rings of tyrosine, phenylalanine and tryptophan become asymmetric as a result of interactions with an asymmetric environment, for example in the folded tertiary structure of a protein, absorption occurring in the near-UV region (320-250nm). These residues can, therefore, act as sensitive reporters of conformational change.

CD is predominantly a solution technique, requires small amounts of sample (100-200μg) and is non-destructive. It can be used to estimate the secondary structure contents of proteins, the effects of environment (e.g. salts or solvents) on conformation, the kinetics of folding or unfolding and thermodynamics (e.g. thermal or chemical denaturation). The disadvantages are that solvents and buffers may interfere in the UV region and only non-absorbing buffers allow measurement below 200nm.

Proteins have several strong absorptions in the IR region which arise from vibrations of the peptide bond. The strongest are the NH stretching vibrations (called amide A) at 3300-3200 cm^{-1}, the amide I band at 1690-1620 cm^{-1}, amide II at 1550-1500 cm^{-1} and the amide III band at 1300-1200 cm^{-1}. These bands are strongly influenced by the conformation of the protein backbone, hydrogen bonding and dipole-dipole interactions between neighbouring peptide bonds cause characteristic bandshifts and bandsplitting, which can be used to identify secondary structure elements. The main problem is that proteins contain a mixture of different structures with similar bands that have considerable overlap. There are two methods to unravel the spectra; mathematical resolution enhancement (deconvolution, band fitting), or pattern recognition techniques, both of which can give estimates of secondary structure contents with errors of a few % (typically 2-5). The advantage of FTIR over CD is its ability to measure, in addition to solutions, concentrated samples, pastes and solids. This is particularly important in systems such as gluten which are intrinsically insoluble.

Infra-Red Spectroscopy

Equipment

The requirements are good baseline stability to avoid artefacts and make spectra comparable and good signal-to-noise ratio (i.e. high-energy throughput) as aqueous samples have strong absorption bands.

Sampling accessories are required depending on the sample, the most useful are:
1. Transmission cells for liquids, with water resistant windows and short path lengths (usually <20µm) because of strong water absorption.
2. Attenuated total reflectance (ATR) cells are the most versatile and can be used for solids, pastes and liquids. Horizontal ATR is the easiest system to load and clean.

Samples

The majority of samples containing gluten proteins are insoluble in water (e.g. dough, gluten and glutenin polymers) while other proteins may also be studied in the hydrated solid state.
1. Solid samples are routinely measured with ATR accessories. The sample is placed on the ATR crystal ensuring good surface contact.
2. An alternative is to immobilise protein films. The sample is placed on an infrared window or ATR crystal, dried down with nitrogen, then re-hydrated.

Isolated proteins (gliadins, LMW or HMW glutenin subunits) available as lyophilised powders can be solubilized in water or D_2O. The addition of acetic acid aids solubility, but information on intermolecular interactions can be lost. The lyophilized protein is stirred in water or D_2O (50mM acetic acid can be added to aid solubility), left to dissolve for several hours, then centrifuged to remove undissolved protein. In the past D_2O has been used to dissolve proteins and peptides, due to problems of subtracting the H_2O spectrum from IR spectra, however, H_2O subtraction is now routine. The protein concentration should be approximately 10 mg/ml.

Spectra Acquisition

Acquisition parameters have to be set correctly to obtain the best results. Good general parameters are: resolution 2 cm^{-1}, 256 scans, spectral range 4000-800 cm^{-1} (other parameters depend on the spectrometer).

Procedure:
1. Measure the single beam spectrum of the dry empty cell as background. (A water-filled cell can be used as background for dilute solutions).
2. Fill the cell with water or buffer solution and measure the absorbance spectrum for solvent subtraction.
3. Empty the cell (flush out with buffer or water) and dry.
4. Fill the cell with sample and measure the spectrum.
5. Clean (flush with ethanol and water) and dry the cell. Proteins tend to stick strongly to crystal surfaces. It may, therefore, be necessary to dismantle the cell for cleaning.

Solvent Subtraction

Before the protein bands can be examined the strong overlaying water spectrum has to be subtracted. The procedure is straightforward for dilute solutions. However, in more concentrated systems the subtraction spectrum has to be multiplied by a weighting factor. Finding the right factor can be a black art! The most commonly used criterion is a flat baseline in those regions where only water absorbs, i.e. below 800 cm^{-1}, at 2500 to 2000 cm^{-1} or around 5500 cm^{-1}. Numerous manual and automatic subtraction procedures have been proposed (Dousseau *et al* 1989) but not all are objective.

It is important to have a matching spectrum to subtract, ideally measured on the same day, as baseline variation can affect results. Buffer salt concentrations and pH must be exactly the same. Buffer salts can also absorb in the infrared and at high concentrations can distort water structure and thus influence the shape of the main water bands.

Resolution Enhancement
1. Derivatives
   The main use is to determine band positions, either from the zero-points of the first derivative, or more easily observable from the negative peaks of the second derivative. All spectrometer software packages

have the option to calculate derivatives. The algorithms may vary between software packages (i.e. Savitsky-Golay, point difference or Fourier derivatives), but usually a degree of smoothing is required to filter the noise. However, very noisy spectra or spectra with water vapour cannot be used. Spectra with water vapour show distinctive sharp rotational bands around 4000 cm^{-1} and 1500-1800 cm^{-1}. It is possible to subtract them from spectra. Some closely overlapping bands cannot be resolved. Also, as the method measures changes in slope, small sharp bands are enhanced more, but wide bands may be lost, so that the area under the band is not constant.

2. Fourier Deconvolution

    Fourier deconvolution has become the standard method for protein structural studies. The principle is that bands with a finite bandwidth transform into a sinus function with an exponential decay directly linked to the bandwidth. Therefore, transforming the spectrum back into the time domain, multiplying the interferogram with an exponential function with the opposite sign, and then transforming into frequency range again, gives stronger bands with smaller half widths. Crucially for structure determinations the area under each band is preserved (Kauppinen *et al* 1981, Griffiths and Pariente 1986).

    The limiting factor is the quality of the original spectrum. At some point on the interferogram the original signal decays to the noise level and beyond that no information can be recovered. It is important to appreciate the limits as at some point only noise and artefacts are amplified.

Band Assignment

Over the years a standard set of secondary structure assignments has been developed. There are some pitfalls, but assignments generally work reasonably well for globular proteins, where they can be directly compared with X-ray or NMR structures (Surewicz *et al* 1993).

In gluten proteins the band assignment is complicated by side chain absorption, in particular glutamine. Glutamine has another effect, in that the amide I band does not usually shift very much (a few wave numbers at most) when H_2O is exchanged for D_2O, but the glutamine side chain amide band can move by up to 25 cm^{-1} from 1658 to 1633 cm^{-1}.

Table 1 gives secondary structure assignments of amide band components in gluten proteins (Pezolet *et al* 1992, Popineau *et al* 1994, Belton *et al* 1995, Wellner *et al* 1996, Gilbert *et al* 2000).

Table 1. Secondary structure assignments of amide band components in gluten proteins.

	Position (cm^{-1})	Structure
Amide I	1690-1695	β sheet and β-turns
	1680-1690	β sheet
	1666	β turns
	1656	α-helix and Gln sidechain
	1640-1650	Random
	1630-1635	β sheet
	1610-1625	Intermolecular β sheet
Amide II	1568	β turns
	1545	α helix
	1524	β sheet
	1517	β sheet and Tyr ring
Amide III	1300	β turns
	1260-1280	α helix
	1220-1245	β sheet

Band Fitting

The relative amounts of each secondary structure can be estimated from the areas under their respective peaks. However, gluten proteins are not very highly structured and the individual components overlap. Consequently, it is not usually possible to fit the original spectra. Even when the amide band has been enhanced, there is often still considerable overlap. Therefore, the spectrum is approximated by a function which is the sum of computer generated absorption bands.

The process

1. Select a set of starting parameters (number of bands, initial band positions, band shapes, approx. band widths and heights, baseline). This can be generated automatically by the software, based on derivatives and residuals. This option is a good starting point, but the results are often not entirely satisfactory and need manual tweaking. An alternative is to enter the parameters in manually.

2. Run the iterative fitting until convergence is reached or the predetermined maximum number of iterations is exceeded. The quality of the fit is indicated by the error function X^2 (the smaller the error the better), or the correlation coefficient.

    The fitting routines use a Levnberg-Marquardt non-linear least squares optimisation algorithm (Marquardt 1963), which will converge rapidly to a minimum, but not necessarily to the global minimum. Some parameters will not fit at all (matrix inversion errors).

3. Check the output - ie mathematically a low x^2 or if the spectrum is 'sensible'
   a. If the fit is OK: go to next step
   b. If the fit is close to convergence tweak the newly generated parameter set and go back to Step 2 to run more iterations.
   c. Sometimes fits are generated which have very low x^2, but are physically unrealistic (i.e. with negative absorption bands, enormously broad bands, unusual baselines etc.). In this case start again from Step 1 using a different set of parameters.
4. Assign each band component to a secondary structure element. Some bands are straightforward, such as the strong 1630 cm^{-1} band for β-sheet. Others may be more problematic. There may be more than one contribution, as is the case for the 1655 cm^{-1} band in solid gluten proteins which can be assigned to α-helix (1660-1650 cm^{-1}) as well as to the glutamine side chain (1658 cm^{-1}). Smaller components in the region of 1700 to 1680 cm^{-1} may arise from either β-sheet or turns.
5. For each structure element combine all components in the amide I region. The secondary structure is given as a % of the total amide I area ((band area/sum of all bands in that region) × 100). As discussed above, the assignment of certain bands, especially high frequency components (sheet, turn) is not clear-cut. This can generate considerable error margins.

Statistical analysis of the amide band shape (pattern recognition) (Lee *et al* 1990, Sarver and Krueger 1991) is possible, but has not been used for gluten proteins. This is mainly due to the lack of an appropriate calibration. The general calibration set is based on globular proteins and may not be applicable, also there are no crystal structures available for calibration.

Examples

Figures 1-3 show the ATR spectra of HMW subunit 1Dx5, the resolution enhancement and bandfit process. Figure 1 shows the spectra of the HMW subunit, water and the protein spectrum after water subtraction. Figure 2 shows the FT-IR ATR spectrum; the second derivative of the spectrum and the deconvoluted spectrum and Fig. 3 the bandfit to the deconvoluted spectrum. The peak areas under the amide I band can then be calculated to determine the secondary structure content of the protein (Table 2).

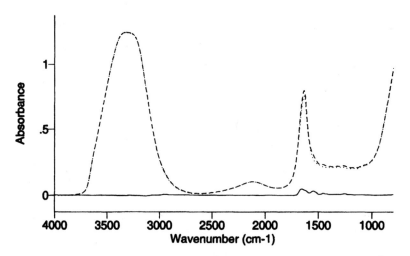

Fig. 1. FTIR ATR spectra of HMW subunit 1Dx5. Resolution 2 cm^{-1}, 256 scans, reference empty cell (ZnSe crystal); HMW subunit 1Dx5, 10 mg freeze dried protein stirred in distilled water (dashed), distilled water (dotted) and protein spectrum after water subtraction (subtraction factor 0.99975)(solid).

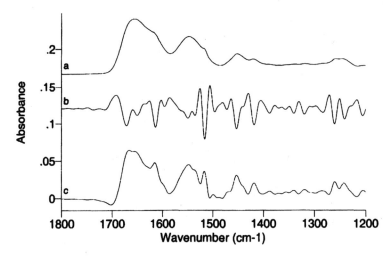

Fig. 2. Resolution enhancement
a) FTIR ATR spectrum in the region 1800-1200 cm^{-1}.
b) Second derivative of spectrum a, Savitsky-Golay, 23 points.
c) Fourier-deconvoluted spectrum a, bandwidth 18 cm^{-1}, resolution enhancement factor 2.0.

165

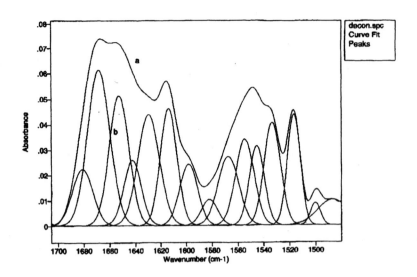

Fig. 3. Bandfit and secondary structure. Amide I and II band regions in the Fourier deconvoluted spectrum from Fig. 2.
a) Original spectrum overlaid with calculated spectrum
b) Band fitted components

Table 2. Fitted components of the amide I from Fig. 3b

Position	Relative Area (%)	Assignment
1681.3	7.5	β-sheet
1668.5	26.1	β-turn
1652.4	18.9	α-helix, gln
1642.0	8.1	β-turn, random
1629.5	17.0	β-sheet
1613.8	14.7	Intermolecular β-sheet
1598.2	7.7	NH$_2$ scissoring of gln

Circular Dichroism Spectroscopy

Equipment

The most commonly used commercial instruments are JASCO and Jobin Yvon machines. Both instruments have software packages for data acquisition, manipulation and structure analysis.

Samples

The proteins should be as pure as possible, as any other proteins (or contaminants) will contribute to the spectrum. The nature and concentration of the buffer can affect the quality of the spectra, any compound that absorbs in the region 190-250 nm should be avoided. Scattering particles, which can contribute to noise, can be removed by filtration through a 0.22 μm filter.

Far-UV CD requires a protein concentration of approx. 0.5 mg/ml, a cell of 0.2 mm path length and a volume of between 50 and 250 μl. The concentration may need to be adjusted to produce the optimum spectra or 0.5 mm path length cells can be used. Smaller path length cells can overcome some of the problems of solvent or buffer transparency. CD instruments cannot make accurate measurements if the combined absorbance of the sample, cell and buffer is greater than 1.0. The absorption spectrum of the sample can be run to determine if there will be transparency problems and the pathlength and/or concentration adjusted. The signal strength is much weaker in the near-UV region than in the far-UV region, and requires longer path lengths (5 or 10 mm) for the same concentrations and approx. 1 ml of protein solution.

Gliadins and glutenins are insoluble in most aqueous buffers, but soluble in aqueous alcohols (70% (v/v) aq. ethanol, 50% (v/v) aq. propan-1-ol) and dilute organic acids (acetic acid) which show low absorption in the far-UV region. Urea and other denaturants absorb strongly below about 220 nm (dependent on concentration), and their use is limited to diagnostic wavelengths above this value.

Spectra Acquisition

Acquisition parameters need to be set to obtain the best resolution. Initially the scan speed can be set high to decide if concentration and sensitivity are correct. The scan speed can then be reduced and the number of scans decided on (increasing scan number increases the S:N ratio) to obtain optimum spectra. Examples of initial and optimised settings are given in Table 3.

Table 3. Examples of initial and optimised settings for CD spectroscopy of gluten proteins.

	Initial settings	Optimised
Wavelength (nm)	180-260	180-260
Scan speed (nm/min)	50	10
Response (sec)	2	4
No. of scans	1	4
Spectral width (nm)	1	1
Resolution (nm)	1	1

The baseline spectrum is measured with water or buffer solution for buffer subtraction. The cell is washed with water and then ethanol and dried over a vacuum line. As the path lengths are small it is important to remove residual solvent/water as these can affect the concentration and buffer composition. The machine must be flushed with pure dry nitrogen before switching on the instrument and while making measurements. Insufficient nitrogen flow can affect the wavelength range that can be scanned, particularly below 200 nm.

Units of CD

The units used in CD can cause some confusion. The JASCO and Jobin Yvon instruments give their data output in millidegrees ellipticity (mdeg), θ. To compare the ellipticity values it is necessary to convert to a normalised value. The instruments will perform the conversions. Two units are commonly used in protein and peptide work, molar ellipticity $[\theta]$ or mean residue ellipticity $[\theta]_{MRW}$.

Molar ellipticity, requires knowledge of the molecular weight of the protein:

$[\theta] = \theta / (10 \times c \times l)$

where c is the molar concentration of the sample and l is the path length of the cell in cm. The units are deg.cm^2.dmol^{-1}. The molar ellipticity $[\theta]$ is related to the difference in molar extinction coefficients by $[\theta] = 3298\Delta\epsilon$.

Mean residue ellipticity requires the mean residue molecular weight and reflects the fact that ellipticity arises from absorption of the peptide bond:

$[\theta]_{MRW} = \theta / (10 \times c_r \times l)$

where c_r is the mean residue molar concentration. The units are the same as for molar ellipticity.

For deconvolution it is important that the inputs are in the correct units for the different software packages.

Determination of Protein Secondary Structure

To obtain structural information from the far-UV (190-250 nm) absorbance the spectrum needs to be compared quantitatively with reference spectra. It is assumed that:

1. The contribution from each of the structural elements is additive
2. The peptide backbone chromophore is the major contributor to the far-UV spectrum (n.b. except in proteins where there are significant numbers of aromatic residues eg. tyrosine residues in HMW glutenin subunits).
3. Each structural element can be described by a single CD spectrum (this is problematic for some gluten proteins that contain low contents of helix and sheet and high contents of turns, of different types, which exhibit different spectra)

Among the earliest approaches was to compare the protein spectra with those of model polypeptides of known structure (Brahms and Brahms 1980) (Fig. 4) or with reference spectra of globular proteins of known secondary structure (Saxena and Wetlaufer 1971). These methods had disadvantages in that there were errors in the reference spectra and some structures could not be unequivocally defined (such as random coil). The methods worked well for helical prediction but were poor for other structural types. More recent methods are based on direct analysis of the CD spectrum, as a simple additive combination of the CD spectra of reference proteins of known secondary structure (for example Provencher and Glockner, 1981; Pancoska and Keiderling, 1991). It is best practice to use the average of the results from several different methods, to obtain a consensus. Websites from which a range of deconvolution programmes can be downloaded are given the Appendix. Problems with prolamins arise from their low (often) contents of helical and sheet structures and that reference proteins are usually globular. Both factors can result in over or underestimates of structure content.

Near-UV Spectra

The near-UV region (250-350nm) is dominated by phenylalanine, tyrosine, tryptophan and disulphide chromophores. If the residues are held rigidly in an asymmetric environment, they will exhibit CD, their CD being sensitive to the overall tertiary structure of the protein. Phenylalanine exhibits bands in the region 250-270nm, tyrosine 270-290nm, trytophan 280-300nm and disulphide bonds exhibit broad but weak bands throughout the neat-UV region. The near-UV region can, therefore, be sensitive to small changes in protein tertiary structure and used to report on structural changes induces by denaturation or heat, for example.

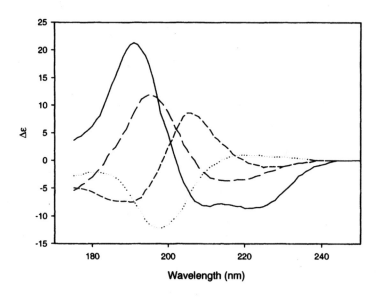

Fig. 4. Circular dichroism spectra of secondary structure types; α-helix
(———), β-sheet (– – –), β-turn (– – – –) and other (⋯⋯⋯). Redrawn from
Brahms and Brahms (1900).

Examples

Examples of reference spectra for secondary structure types based on
model polypeptides are shown in Fig. 4 while Table 4 summarises the
positions of positive and negative bands.

Table 4. Positions of positive and negative bands associated with different
secondary structures.

α-helix	positive at 190-195nm
	negative at 206nm
	negative at 222nm
β-sheet	negative at 215-220 nm
	positive at 195-200nm
Random coil	positive at c.220nm
	negative at c.200nm
β -turns	do not exhibit a dominant type of spectrum,
	exhibit a range of spectral types

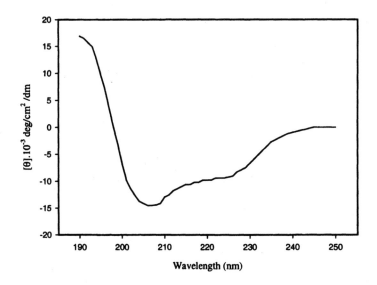

Fig. 5.   CD spectrum of A-gliadin in water, pH5 adjusted with HCl. Redrawn from Purcell *et al* (1988).

Figure 5 shows the spectrum of A-gliadin (redrawn from Purcell *et al* 1988). The spectrum is similar to that of the α-helical spectrum in Fig. 4, with negative bands around 208-209nm and 221-222 nm. Deconvolution using the method of Chang *et al* (1978) gave values of 28% α-helix; 38% β-structure and 34% unordered structures (including β-turns).

Appendix
Free deconvolution software and further background to CD can be found at www.umdmj.edu/cdrwjweb and www.imb-jena.de/ImgLibDoc/cd/index

Acknowledgement

Long Ashton Research Station and IFR receive grant-aided support from the Biotechnology and Biological Sciences Research Council of the United Kingdom.

References

Belton, P.S., Colquhoun, I.J., Grant, A., Wellner, N., Field, J.M., Shewry, P.R. and Tatham, A.S. 1995. FTIR and NMR studies on the hydration of a high M_r subunit of glutenin. Int. J. Biol. Macromol. 17:74-80.
Brahms, S. and Brahms, J. 1980. Determination of protein secondary structure in solution by vacuum ultraviolet circular dichroism. J. Mol. Biol. 138:149-178.

Dousseau, F., Therrin, M. and Pezolet, M. 1989. On the spetral subtraction of water from the FTIR spectra of aqueous solutions of proteins. Appl. Spectros. 43:538-542.

Gilbert, S.M., Wellner, N, Belton, P.S., Greenfield, J.A., Siligardi, G., Shewry, P.R. and Tatham, A.S. 2000. Expression and characterisation of a highly repetitive peptide derived from a wheat seed storage protein. Biochim. Biophys. Acta 1479:135-146.

Griffiths, P.R. and Pariente, G.L. 1986. Introduction to spectral deconvolution. Trac-Trends Anal. Chem. 5:209-215.

Kauppinen, J.K., Moffat, D.J., Mantsch, H.H. and Cameron, D.G. 1981. Fourier self-deconvolution - a method for resolving intrinsically overlapped bands. Appl. Spectros. 35:271-276.

Lee, D.C., Harris, P.I., Chapman, D. and Mitchell, P.C. 1990. Determination of protein secondary structure using factor analysis of infrared spectra. Biochem. 29:9185-9193.

Marquardt, D.W. 1963. An algorithm for the estimation of non-linear parameters. J. Soc. Ind. Appl. Math. 11:431-441.

Pancoska, P. and Keiderling, T.A. 1991. Systematic comparison of statistical analyses of electronic and vibrational circular-dichroism for secondary structure prediction of selected proteins. Biochem. USA 30:6885-6895.

Pezolet, M., Bonenfant, S., Dousseau, F. and Popineau, Y. 1992. Conformation of wheat gluten proteins - comparison between functional and solution states as determined by infrared spectroscopy. FEBS Lett. 299:247-250.

Popineau, Y., Bonenfant, S., Cornec, M. and Pezolet, M. 1994. A study by infrared spectroscopy of the conformations of gluten proteins differing in their gliadin and glutenin compositions. J. Cereal Sci. 20:15-20.

Provencher, S.W. and Glöckner, J. 1981. Estimation of globular protein secondary structure from circular dichroism. Biochemistry 20:33-42.

Purcell, J.M., Kasarda, D.D. and Wu, C.-S.C. 1988. Secondary structures of wheat α- and ω-gliadin proteins: Fourier transform infrared spectroscopy. J Cereal Sci. 7:21-32.

Sarver, R.W. and Krueger, W.C. 1991. Protein secondary structure from Fourier transform infrared spectroscopy: A data base analysis. Anal. Biochem. 194:89-100.

Sarver, R.W. and Krueger, W.C. 1991. An infrared and circular dichroism combined approach to the analysis of protein secondary structure. Anal. Biochem. 199:61-67.

Saxena, V.P. and Wetlaufer, D.B. 1971. A new basis for interpreting the circular dichroic spectra of proteins. Proc. Natl. Acad. Sci. USA 68:969-972.

Surewicz, W.K., Mantsch, H.H. and Chapman, D. 1993. Determination of protein secondary structure by Fourier transform infrared spectroscopy: a critical assessment. Biochem. 32:389-394.

Wellner, N., Belton, P.S. and Tatham, A.S. 1996. Fourier transform IR spectroscopic study of hydration induced structure changes in the solid state of ω-gliadins. Biochem. J. 319:741-747.

Chapter 9

Small-Scale Quality Measurements

F. Békés
CSIRO Plant Industry, PO Box 1600, Canberra, ACT 2601, Australia

O.M. Lukow
Cereal Research Centre, Agriculture and Agri-Food Canada,
195 Dafoe Road, Winnipeg, Manitoba, R3T 2M9

S. Uthayakumaran
Cereal Research Centre, Agriculture and Agri-Food Canada,
195 Dafoe Road, Winnipeg, Manitoba, R3T 2M9

G. Mann
School of Food Biosciences, The University of Reading, PO BOX 226,
Whiteknights, Reading RG6 6AP, Berkshire, UK

Introduction

The development of small-scale dough testing equipment and the associated automated interpretation of the resulting mixing curves has provided better reproducibility and removed operator bias, resulting in more objective assessment of the experimental variables. This has facilitated a wide range of research and several applications in breeding in which either only limited amounts of test material have been available or the more objective, precise assessment of data offered extra benefits. (For the different applications of small scale dough testing methods as research and breeders tool see the reviews of Békés and Gras 1999, 2000 and Gras *et al* 2001). Small- and micro-scale testing procedures such as the two-gram Mixograph, micro-extension tester and micro-baking methods (Rath *et al* 1990; Gras *et al* 1990; Rath *et al* 1994; Gras and Békés, 1996; Ingelin and Lukow, 1999; Suchy *et al* 2000, Uthayakumaran *et al* 2000a,b) were originally developed to mimic the traditional scale methods using significantly smaller sample sizes. The development of equipment and procedures therefore always included the validation against the traditional

173

basic procedures, such as AACC or ICC standard methods. The justification of using these methodologies is the reasonable, sometimes excellent relationships between the traditional and small-scale methods.

Mixograph Tests

The Mixograph is the basic instrument to study dough development and the rheology of the resulting dough in a process where the mixing is carried out using standing and rotating pins to provide the mixing action. The mathematical interpretation of the mixing curve of the Mixograph indicates that the mixing action in a pin mixer is the superposition of a huge number of strechings and foldings of the dough among the pins (Gras *et al* 2001). Parameters derived from the mixing curve – provided by the different computer softwares developed recently for the small-scale machinery – are the measures of the mixing requirement, dough strength and dough stability, they can even be related to the extensional behavior of the dough. Several small-scale Mixographs have been developed and used in different laboratories (Voisey *et al* 1969; Gras *et al* 1990), requiring 2 g and 10 g of flour. The basic procedures applied on these equipments are analogous, and follow the steps given here.

Apparatus
1. Mixograph with 10 g water jacketed mixing bowl (maintained at 25°C), pins, accessories (Voisey *et al* 1969), speed fixed at 92 rpm; or
   2 g Mixograph^{TM} (National TMCO, Lincoln, NE, USA); or
   10 g Mixograph^{TM} (National TMCO, Lincoln, NE, USA); or
   ReoMixer (Reologica Instruments AB, Yardley, PA).
2. Computer (programmed for data collection) or chart recorder
3. Water bath to maintain the temperature (at approximately 30°C) of the distilled water used for the test
4. Precision balance (readability of 0.01 g)

Procedure
Two basic procedures are applied depending on the amount of water used in the mixing experiment:
- Constant water procedure where flour is weighed into the mixing bowl and constant amount of distilled water (6.2 mL for 10 g of flour) is dispensed into the 10 g mixing bowl containing the flour.
- Using the method, analogous to the AACC approved method 54-40 (AACC, 2000), the amounts of flour (14% moisture basis) and water to be used for mixing are calculated from the protein and moisture contents of the flour. The equation is given as:
Optimum water absorption (14% moisture basis) = 1.5 percent flour protein (14% moisture basis) + 43.6

The mixing bowl is immediately placed in position on the Mixograph and the mixogram recording started. Before starting each mixogram, the sample number, data and other relevant information are entered into the computer. Mixograms can be recorded electronically using a computer, and software is available for data analysis (National Mixsmart program, Pon *et al* 1989).

Recording of mixing curve can also be made on a chart recorder (Approved method 54-40, AACC, 2000).

Calibration of Mixograph needs to be performed daily prior to use. This is carried out by adjusting the torque baseline to zero while running the Mixograph empty.

Mixing Parameters Determined on the Mixograph

Several computer software packages are available in general for interpretation of the mixing curve:
1. The software for a multi-channel, computer-based system for analyzing dough rheology developed by Pon *et al* (1989)
2. The Mixsmart program, commercially available with the equipment manufactured by National TMCO, Lincoln, NE, USA)
3. The MD06 developed by Gras *et al* (1990)

Some of the most important parameters determined using these computer programs are shown on Fig. 1.

Regardless of which computer program is used, the most important features of the Mixograph curves are the same:
- the × coordinate of the maximum of the midline, (Mixograph Development Time, MDT) or Mixing Time, MT, measured in min, the time measured for the dough to reach maximum resistance on midline;
- the y coordinate of the maximum of the midline, (Peak height, PHG, or Peak Resistance, PR) measured in Mixograph torque units (N.m or % of torque scale), the height from baseline to the center of the curve (midline);
- the slope of decline of resistance after the peak (Resistance Breakdown, RBD), a measure of dough stability, measured as a percentage of PR
- parameters related to the changes of bandwidth during the mixing such as the maximum bandwidth (MBW), the bandwidth at PR (BWPR) and the slope of its decline after the peak (Bandwidth Breakdown, BWBD) These values are directly available with MD6 while using Mixsmart they can be calculated from the parameters of upper and lower envelop curves.

Several data treatments provide information also about the energy consumption during the phases of mixing action such as Energy to Peak (ETP) or Total Energy (TEG), measured in N.m, the energy is determined by area under mean torque curve from the origin to the peak value and whole torque curve, respectively.

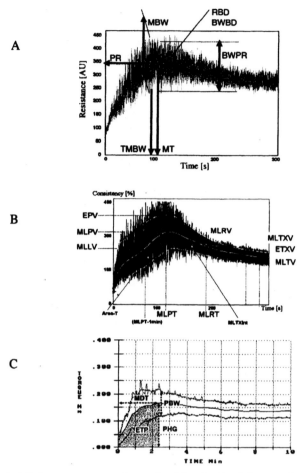

Fig. 1. Mixing parameters determined by the 2 g Mixograph, using (A) the MD6 (Gras *et al* 1990), (B) the Mixsmart software and, (C) the program developed by Pon *et al* (1989). For (A), MT (mixing time), PR (peak resistance), BWPR (bandwidth at peak resistance), RBD (resistance breakdown), BWBD (bandwidth breakdown), TMBW (time to maximum bandwidth and MBW (maximum bandwidth). For (B), EPV (envelope peak value, %), ETXV (envelope time × value, %), MLLV (mid-line left value (%)), MLPT (mid-line peak time, min), MLPV (mid-line peak value, %), MLRT (mid-line right time, min), MLRV (mid-line right value, %), MLTV (mid-line tail value, %), MLTXV (mid-line time × value, %), MLTXint (mid-line time × integral, %TQ*min), Area-T (envelope area, %TQ*min). For (C) MDT (Mixograph development time), ETP (energy to peak), PBW (peak bandwidth), and PHG (peak height).

176

Micro-scale Farinograph Tests

The plasticity and mobility of a dough subjected to a prolonged, relatively gentle, mixing at constant temperature is measured in different Z-arm mixers such as Farinograph or Valorigraph. Apart from the determination of important mixing parameters such as dough development time and dough stability, this test is generally used for evaluating water absorption of flours, applying the linear relationship between the maximum resistance of the dough and the amount of water used in the mixing.

Apparatus
1. Two small-scale Z-arm type mixers are available and used routinely in different laboratories: Brabender Farinograph with a 10 g mixing bowl
2. Micro Z-arm mixer (Gras *et al* 2000) equipped with a 4 g mixing bowl (Fig. 2)
3. Precision balance (readability of 0.01 g)
4. Burette, 25 ± 0.02 mL
5. Water Bath (30°C)

Procedure
Using the Brabender equipment, the bowl temperature is maintained at 30°C. The chart paper needs to run horizontally. Scale head pointer and the writing pen are adjusted to zero. The burette is filled with distilled water and maintained at room temperature. Moisture content of flours is determined as directed in oven method for flour (AACC approved methods

Fig. 2. Prototype micro Z-arm mixer developed by CSIRO, Technical University of Budapest and METEFEM, Budapest, Hungary.

177

44-15A). Ten grams of flour are weighed (at 14% moisture basis) into the 10 g mixing bowl. The Farinograph machine is turned on and run for 1 min until 0 min line is reached. At this point water is added into the bowl (volume based on expected absorption) within 25 s. The bowl is covered with a glass plate to prevent evaporation. If the curve is beyond 500 BU more water is added. The maximum resistance of the curve needs to be centered on 500 ± 10 BU. The length of test run is 20 min. In case of extra strong wheat, the length of the run could be extended up to 40 min.

Only 4 g of flour is needed to use the micro Z-arm mixer. The computer-regulated equipment with automatic water-pump system and computer software provides a procedure to determine water absorption from one single mixing experiment using an algorithm based on the following equation:

$$WA\% = ML/0.04 + [(PR-500)/(125{,}58-(1{,}445*ML/0.04))],$$

Where, WA - the calculated water absorption
ML - the amount of water added to 4.00 g of flour
PR - peak resistance found using ML amount of water

Mixing Parameters Determined on Micro-Farinographs

Figure 3 illustrates the data obtained using both types of equipment. The parameters determined using either manual or electronic data processing procedures are the following:
- Farinograph water absorption, amount of water added to the flour to reach 500 BU (Brabender Units) dough consistency;

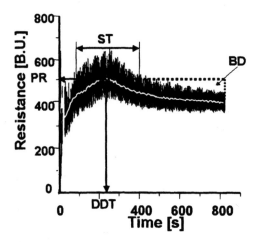

Fig. 3. Parameters determined on the Micro Z-arm mixer: DDT – dough development time, PR - peak resistance, ST – stability, BD – breakdown. Resistance is measured in Brabender Units (BU).

- Dough development time (DDT): According to AACC Method 54-21, it is the interval to the nearest 0.5 min, from first addition of water to the point in maximum consistency range immediately before first indication of weakening. Occasionally two peaks may be observed and the second peak should be taken for determination of dough development time;
- Stability (ST): It is defined as time difference to closest 0.5 min, between the points where the top of curve first intersects 500 BU and where it leaves 500 BU line;
- Breakdown (AACC 54-21) (BD): Difference in BU from top of the curve at peak to top of the curve 5 min after peak.

Small-scale Extension Test

The Extensigraph offers an empirical measurement of stress-strain relationships in a dough with defined rest periods and dough geometry (Preston and Hoseney, 1991). Extensigrams are used to assess general flour quality such as resistance to stretching and extensibility (dough extension properties). They also measure dough strength and stability and give the baker an idea how the dough would behave in his baking equipment.

Two pieces of equipment are available for small-scale extension testing: The TA-XT2 Texture analyzer with the special Kieffer rig (Kieffer *et al* 1981) and the Micro-extension tester (Rath *et al* 1994). The procedures recommended by the developers differ first of all in the speed of stretching; the final results provided by the two different types of equipment however are comparable (Gras, unpublished results). On the new version of the Micro extension-tester, developed recently, the speed of stretching can be altered over a large interval (0.2-10 mm/sec). This machine and the related computer program allow the user to define any speed profile during the stretching, so experiments with accelerating speed or with conditions emulating any other equipment can be carried out.

Method for TA-XT2 Texture Analyzer
Depending on the availability of the raw material, there are two different procedures using either 2 g or 10 g flour.

Procedure a

Apparatus
1. Texture Technologies Corp., (Scarsdale, NY), with a Kieffer rig
2. 2 g Mixograph (National TMCO, Lincoln, NE) or equivalent
3. Teflon-coated block (Kieffer *et al* 1981) and teflon strips (2 × 60 mm)
4. A temperature controlled chamber
5. Analytical balance (readability 0.01 g)

Two grams of flour are weighed (at 14% moisture basis) into the 2 g mixing bowl. Salt solution (to give 2% of the flour mass) and water (flour water absorption (FAB)+5 (calculated for 2 g flour) minus liquid contributed by the salt solution) is added to the flour. The mixing bowl is immediately placed in position on the Mixograph and the mixogram recordings are started. Dough is mixed to peak dough development time, taken out from the mixing bowl and rounded into a ball and placed over three or four channels of the teflon-coated block. Prior to the placement of dough, the teflon-coated block is prepared by placing non-adhesive teflon strips (2 × 60 mm) coated in mineral oil in the channels. Once the dough is placed in the teflon-coated block, the upper half of the block is placed in position and tightly clamped, which distributes the dough over two to three channels, to yield dough strips of uniform geometry. The dough is rested for 40 min at 25°C prior to the test. The dough strips are separated from the teflon strips, positioned across the Kieffer rig dough holder and immediately tested on the Texture Analyzer at a hook speed of 3.3 mm/sec and a trigger force of 1 g.

Procedure b

Apparatus
1. Stable Micro Systems (Surrey, UK), with a Kieffer rig
2. 10 g Mixograph (Simon Mixographic mixer)
3. Teflon-coated block (Kieffer *et al* 1981) and teflon strips (17x60 mm)
4. A proofing oven
5. Analytical balance

The procedure applying the Kieffer rig was developed by Stable Micro Systems (SMS) and used with the TA.XT2 texture analyzer. To produce the dough for the experiment, 10 ± 0.1 g of flour is transferred into the 10 g mixing bowl of a Simon Mixographic mixer and mixed for 30 s at 60 revolutions per min (rpm). Water is then added at the water absorption level determined by the CCFRA method (Campden and Chorleywood Food Research Association, UK) and mixing continued to the optimum dough development time (as determined by 2 g Mixograph). The dough is then removed and rolled gently into a ball. The dough ball is placed into a small plastic pot (60mL) with the cap on and left to relax for 20 min at ambient (temperature controlled room at 20°C). In the meantime the lametta strips are placed in the grooves of the base of the molder and lubricated with paraffin oil so that the samples can be easily removed without any stress introduced into the dough strips.

After relaxation the dough is rolled into a ball and then into a sausage shape (five rotations to both sides in between the palms of both hands). The sausage shape dough is placed on the grooved base with its length perpendicular to the groove direction. The top is placed over the bottom from above and pressed till the excess dough is extruded from the sides. The excess dough is then removed with a spatula. This form cuts the samples into strips and allows the dough to relax, whilst preventing moisture loss. The molder is placed in the clamp and screwed down. Excess dough extruded on clamping is removed again with a spatula. This ensures the separation of each strip (any connected strips would be damaged by separation after they have undergone relaxation). Finally, the dough clamp (with the molder) is placed to relax in the proofing oven with > 85 RH (relative humidity) at 30°C for 40 min.

After the relaxation time the tension in the clamp is released and the molder slid out. The top part of the molder is slid backwards gently over the grooved base, revealing each dough strip one at a time. Each dough strip is taken with a small spatula and placed across the grooved region of the sample plate and the lametta strip is removed gently. The plate is inserted into the rig by holding the spring-loaded clamp and the test started. Once a trigger force of 5 g is attained, the hook extends the dough sample centrally until its elastic limit is exceeded and the dough is deformed (Smewing 1995; Mann, 2001).

Recommended Settings for micro extension tests on dough or gluten, using the TA.XT2 Texture Analyzer with a 5 kg load are:

Mode	Measure Force in Tension
Option	Return to Start
Pre-test Speed	2.0 mm/s
Test Speed	3.3 mm/s
Post-test Speed	10.0 mm/s
Distance	*180 mm
Trigger force	Auto- 5 g
Data Acquisition Rate	200 pps

* Adjusted accordingly depending upon the maximum extension distance.

Method for the Micro-extension Tester

Apparatus
1. Micro-extension tester (Fig. 4)
2. 2 g Mixograph™ (National TMCO, Lincoln, NE)
3. Prototype molder
4. Beaker, syringe, blue towel pieces, cotton swabs
5. Humidity chamber
6. Analytical balance (readability of 0.001 g)

Procedures
The total amounts of flour (2 g, 14% moisture basis) and water to be used for extension testing are calculated from the protein and the moisture

contents of the flour using the standard method (Method 54-40, AACC, 2000). Flour, salt solution (salt, 2% of the flour mass) and water (required water minus liquid contributed by salt solution) are transferred into a 2 g mixing bowl and dough for extension testing is mixed to peak dough development. Dough is taken from the mixing bowl and 2 dough pieces (1.7 g) are cut into cylinders (approximately 6 mm in diameter) with a prototype molder consisting of a 153 mm diameter drum rotating at 20 rpm within a fixed 167 mm diameter partial outer drum. Dough pieces (1.7 g/test) are introduced and rolled around the annular space for approximately 300°. The molded pieces, approximately 45 mm in length are mounted on a sample carrier and rested at 30°C and >90% RH for 45 min before extension testing (Gras and Békés, 1996). Extension is performed in a micro extension tester, with a 19 mm gap and 6 mm hook operating at 10 mm/s. Recordings of the dough resistance and the sample carrier position are taken at 100 points/s and recorded by a personal computer, using LabTech Notebook software.

Note: The Approved AACC method to determine extensibility uses a Z-arm type of mixer to make the dough for the extension measurement. The small-scale extension test can also be carried out on this type of mixer (Haraszi *et al* 2001).

Commercialization of the micro-extension tester and prototype molder is in progress.

Parameters Determined with the Small-Scale Extension Tests

The parameters determined from the extension curves using the software provided with the two types of equipment are identical (Fig. 5).

1. Maximum resistance ($R_{max,}$, g), is the maximum force needed to rupture the dough. This is related to the elastic character of the dough
2. Extensibility (Ext, mm), the distance from the beginning of the test till the point at which the dough ruptures. This is related to the viscous character of the dough
3. Viscoelastic ratio determined by the ratio of maximum resistance to extension (R_{max}/Ext)
4. Area to peak, area under curve till R_{max}, which is related to the energy needed to deform the dough

Recently a new measurement on the Micro-extension Tester was introduced (R. Anderssen *et al* unpublished). It was found that better reproducibility could be achieved if the distance to maximum resistance was used instead of the point of rupture for the interpretation of extensional properties. The updated version of the Micro-extension Tester software facilitates measurement of this new parameter.

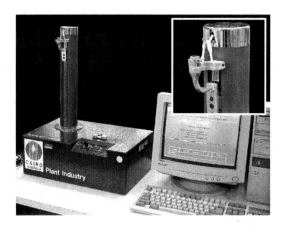

Fig. 4. Prototype Micro Extension tester developed by CSIRO, Australia.

A

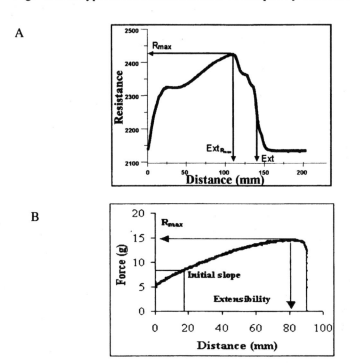

B

Fig. 5. Typical extension curve and parameters determined by the Micro Extension Tester (A) and by the Kieffer rig used with the TA.XT2 texture analyzer (B). The differences in the details of the curves are caused by differences in the speed of stretching used on the two types of apparatus (10 and 3.3 mm/s, respectively).

End-product Quality Tests

The ultimate tests of wheat quality are where the flour is evaluated for its ability to make good quality end products. End product quality tests mimic different industrial technologies with the end products being scored to assess quality. This procedure is a complex and difficult task using the traditional procedures and sample sizes. The exercise is even more problematic when the sample size is reduced. Small-scale end-product quality testing is therefore only a modelling procedure, the aim is not to precisely describe quality but to provide trends, ranks, cut-off values. Having considered the limitations of these methodologies, they can be extremely useful as research and breeding tools where trends, ranks and cut-off values can be defined or cause/effect relationships can be established.

The relative importance of the target product and target technology to be mimicked by small-scale end-product quality tests is different from country to country. Some specific examples are given here illustrating how the quality of different bread- and noodle-types can be estimated on small-scale (10-20 g flour), even micro-scale (below 5 g flour) levels.

Small-Scale Baking Tests

Apparatus
1. 35 g Mixer (National TMCO, Lincoln, NE or equivalent) with a recommended Orbital speed of 92 rpm
2. Chart recorder, Goertz Servogor 120 and a modified energy input meter (Kilborn and Tipples 1981)
3. Cabinet for fermentation and proofing dough set at 37.5°C and 85% RH, (National TMCO, Lincoln, NE or equivalent);
4. Sheeter with 6-inch rolls (National Mixograph, National Manufacturing Co., Lincoln, NE or equivalent)
5. Baking pans constructed of 2X tin with the following dimensions:

	Length (cm)	Width (cm)	Depth (cm)
Top	8.7	4.2	4.0
Bottom	7.0	3.3	4.0

6. Oven, rotary, (National TMCO, Lincoln, NE or equivalent) with capability of maintaining temperature of 220 ± 8°C. Moisture should be high enough so that moisture film is seen on dough after 10 s of being in the oven
7. Fermentation bowls (plastic)
8. Timer for mixer
9. Instrument to measure loaf height
10. Loaf volumeter, rape seed-displacement type
11. Balance, scoops, spatula, pipettes, graduated cylinders, beakers etc.

Procedure

Two different procedures commonly used to prepare small size loaves are described here.

1. The Canada Short Process (CSP), a no-time straight dough process developed at the Grain Research Laboratory, Winnipeg, Canada (Preston *et al* 1982).
2. AACC Optimized Straight-Dough Breadmaking Method (Method 10-10, AACC 2000).

The Canada Short Process (CSP)

Small size loaves are baked using 35 g of flour (14% moisture basis). The formulation is given as a percentage of flour mass. Flour (100%), sugar (4%), salt (2.4%), whey powder (4%), shortening (3%), malt (0.6%), instant dry yeast (2%), ammonium phosphate (0.1%), L-ascorbic acid (150 ppm) and water (FAB minus liquid contributed by ammonium phosphate, ascorbic acid and malt solution) are mixed until 10% past peak dough development time in a 35 g Mixograph. The dough is removed from the mixing bowl; the weight recorded and placed in a lightly greased numbered fermentation bowl. The bowl is covered with a lid and placed in the 37.5°C proofing cabinet. The punching and panning times are filled out in a baking schedule. After resting for 1 min, the dough piece is removed from the bowl and punched/rounded lightly seven times by hand. The dough is placed back in the covered bowl and placed in the 37.5°C proofing cabinet for a further 15 min. The rounded dough piece is removed from the fermentation bowl 2 min before the designated panning time. The dough piece is passed through the sheeting rolls 3 times, with progressively smaller gap widths (1st pass – 11/32" gap, 2nd pass – 3/16" gap, 3rd pass – 1/8" gap), and molded. The molded dough is placed in the pan (dough seam down towards the side of the pan that is greased only half way up) and the ends are pushed downwards with the tips of the fingers producing a pillow shape appearance. Panned dough is placed in the proofing cabinet and proofed to a height of 60 mm and baked for 15 min at 220°C. The loaf is then removed from the oven and the loaf height in the pan is recorded. Loaf is removed from the pan and set on cooling rack.

AACC Optimized Straight-dough Breadmaking Method

This method provides a basic baking test for evaluating bread wheat flour quality by a straight-dough process that employs long fermentation and in which all ingredients are incorporated in the initial mixing step. The method is a 180 min sugar-based fermented dough system with shortening. Small size loaves are based on 35g of flour (14% moisture basis). Flour (100%), sugar (6%), salt (1.5%), whey powder (4%), shortening (3%), instant dry yeast (1%), ammonium phosphate (0.1%), L-ascorbic acid (20 ppm), malt (0.2%) and water (FAB - 3 minus liquid contributed by ammonium phosphate, ascorbic acid and malt solution) are mixed until 10%

past peak dough development time in a 35 g Mixograph. The dough is removed from the mixing bowl; dough weight recorded and is placed in a lightly greased numbered fermentation bowl. The bowl is covered with a lid and placed in the 37.5°C proofing cabinet. The punching and panning times on the baking schedule are filled out. The dough is removed from the fermentation bowl, and run through the sheeter at 3/16, it is folded twice and placed in the fermentation bowl, and returned to the fermentation cabinet for 50 min. The rounded dough piece is removed from the fermentation bowl 2 min before the designated punch time and run through the sheeter at 3/16. The dough is folded twice, and placed back in the fermentation cabinet for another 25 min. After 25 min the dough is taken from the fermentation cabinet and passed 3 times through the sheeter at: 11/32; 3/16; and 1/8 and molded. The molded dough is placed in the pan (dough seam down towards the side of the pan that is greased only half way up) and the ends are pushed downwards with the tips of the fingers producing a pillow shape appearance. The sample is proofed to a height of 60 mm then baked for 15 min at 220°C. The loaf is then removed from the oven and the height of the loaf in the pan is recorded. Loaf is removed from the pan and set on cooling rack. After cooling for 30 min, the loaves are weighed and the loaf volume is measured and recorded on the baking sheet. The loaves are placed in plastic bags for further evaluation, which is carried out after 24 hours.

Evaluation of Bread Quality

Loaf volume

The loaf volume (LV) is measured using the rapeseed displacement volumeter.

Bread scoring

Appearance is scored on a scale of 1 (poorest) to 10 (best). This score takes into consideration the degree of break-and-shred, loaf symmetry, loaf bottom and crust appearance. A good quality loaf has a high break-and-shred, a flat bottom, even shape and even crust color.

Crumb structure is scored on a scale of 1 (poorest) to 10 (best). The loaves are cut in half longitudinally and the two crumb sides are examined. The first feature to be considered is the cell wall thickness and the score is reduced if the crumb is very open, holey, streaked or uneven. High quality bread has thin cell walls, and a fine, even cell structure.

Photography

The bread is photographed with a digital camera to give a permanent record. Photographs can also be used when publishing the data.

Bread crumb grain

Evaluation of bread crumb grain is carried out using the CrumbScan software (American Institute of Baking, Manhattan, KS) where the images of the two slices of freshly cut bread are scanned and analyzed for cell size and shape and reported as composite finess (CF).

Crumb color

Crumb color is measured using the Minolta Chromameter (Model CM 525i, Minolta, Japan) and the brightness (L*), redness (a*) and yellowness (b*) are recorded.

Crumb firmness

The crumb firmness (CRF) is evaluated by a compression (according to manufacturer's instructions) test on two stacked bread slices using the Texture Analyzer and reported as force (g) at 25% and 40% compression. The probe used is the TA-4 1.5" (38 mm) acrylic probe with a chamfer. The parameters obtained are:
1. Force required to compress the crumb to 25% and 40% (g), modified AACC approved method 74-09 (AACC 2000)
2. Maximum force or peak force applied during the test

Micro-Scale Baking Tests

Different versions of micro-scale baking tests exist, but only two procedures are described here: straight and sponge dough baking tests.

Apparatus
1. 2 g Mixograph™ (National TMCO, Lincoln, NE)
2. Prototype molder
3. Cabinet for fermentation and proofing dough set at 40°C and 90 % relative humidity
4. Oven, rotary, National Manufacturing Company, with capability of maintaining temperature of 200°C
5. Analytical balance (readability of 0.001 g)
6. Thimbles (17 mm in diameter) (Fig. 6)
7. Timer for mixer
8. Vernier calipers

Procedure

Two procedures are described here adopting rapid and 'sponge and dough' techniques.

The rapid dough method is suitable to evaluate the overall baking performance of flours with average quality.

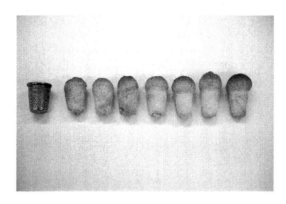

Fig. 6. Micro-baked loaves made from 2 g of flour. The figure indicates the effect of protein content on loaf size: flour of cultivar Banks (D) with 10% protein content was supplemented with starch increased amount of its starch (A, B, and C) and gluten (E, F and G).

Rapid Dough Formulation

	Flour Basis (%)
Flour, 14% moisture basis	100.0
Salt, chemically pure NaCl	2.0
Compressed yeast	2.5
Improver	0.5

The "rapid dough" formulation improver contains 100 ppm of ascorbic acid and 0.5 SKB units of cereal α-amylase per 100g of flour.

The total amounts of flour (2 g, 14% moisture basis) and water to be used calculated from the protein content (determined by the Dumas total combustion or Kjeldal Method) and the moisture content (determined by AACC Methods 44-15A) of the flour using the standard method (Method 54-40, AACC, 2000). The ingredients are mixed to optimum development in the 2-g Mixograph. A 2.4 g dough piece is weighed and molded in a wooden molder placed in a plastic container and rested for 20 min at a temperature of 40°C and a RH of 90%, after which it is remolded and placed in a thimble. The dough placed in the thimble is proofed for 45 min at 40°C and 90% RH and baked at 200°C for 17 min (Gras and Békés, 1996). Loaf height (LH) is measured with vernier calipers.

Geometric considerations show that loaf volume is not directly proportional to loaf height in these small loaves. However, the relation can be viewed as approximately linear over the range of interest.

Sponge and Dough Formulation

The sponge and dough method differentiates strong flours much better than the rapid method (Butow *et al* 2002).

	Flour Basis (%)
Flour, 14% moisture basis	70.0
Compressed yeast	3.0
Malt extract	0.5
$(NH_4)_2SO_4$	0.06
$CaHPO_4$	0.25

The total amounts of flour (2 g, 14% moisture basis) and water to be used are calculated from the protein content (determined by the Dumas total combustion or by Kjeldal Method) and the moisture (determined by AACC Methods 44-15A) of the flour using the standard method (Method 54-40, AACC, 2000).

70% of the calculated amount of the flour is mixed 30 s in the 2 g Mixograph with 65% of the calculated water. Incubate the sponge for 2 hours at 40°C, add the remaining flour and water with 2% salt and mix the dough till optimum then continue the process described above for the rapid method.

Small-Scale Noodle Quality Tests

Noodle products represent a large proportion of the world's wheat production and consumption. Similarly to bread and bread-making quality, numerous noodle products with widely different requirements for flour composition and quality attributes are produced in different countries. For white-salted noodles (WSN) (Ross *et al* 1996), in particular for the Japanese market, it is desirable for the noodles to have soft and elastic (or springy) bite and smooth surface characteristics. To achieve these characteristics flour is selected which has appropriate, medium protein content and in which the starch, or the flour have high paste viscosities. The high swelling starches that are associated with the desirable soft and elastic eating quality of the Japanese WSN have been linked to a null phenotype for the granule bound starch synthase (GBSS) protein on chromosome 4A (Zhao *et al* 1998). Flour protein content also affects WSN texture with high protein content generally being associated with firmer WSN. In general, desirable textural properties for yellow alkaline noodles (YAN) – another important noodle variety - are firmer than those for WSN (Miskelly and Moss 1985, Ross *et al* 1996). YAN should have firm and elastic bite characteristics, and a smooth surface when cooked. Both high protein content and high protein quality have been shown to be associated with firmer bite and increased cutting stress and higher levels of elasticity in YAN. For YAN, high paste viscosities and swelling power have been shown to be negatively related to firmness.

Research on flour components for noodle making is still in its infancy compared to work on breadmaking. Given the importance of this market there is still much to learn about the role of flour components for noodles, and the acquired knowledge can be incorporated into superior wheat varieties. The task is to elucidate the role of individual starch, protein, lipid and non-starch carbohydrate components of flour in relation to noodle quality attributes (Konik *et al* 1992). It requires technology for the production and testing of noodles made from less than 20 g flour that can be applied in breeding, marketing and quality control. Experiments on model flours (Yun *et al* 1998, 1999) can provide the basis for surveys of large populations of breeder's lines where variation in key components can be determined as selection criteria for breeding and as indicators for specific genetic modification of noodle wheat.

A small-scale noodle machine developed recently (Quail *et al* 1999) requires not more than 10 g of flour to produce different noodle types. This equipment together with texture-determination methods (Ross *et al* 1999) allows determination of the noodle making properties of samples available only in small quantities, such as breeders material and flour samples where the chemical composition has been systematically altered. Results of comparing the products of the traditional scale Otake machine with those from the micro-scale equipment showed that the micro scale method is a very effective reproduction of the Otake method (Quail *et al* 1999). This success appears to be due to the design of the micro noodle machine to precisely reproduce the roll diameter and speed of the Otake machine.

Apparatus
1. 10 g Mixograph™ (National TMCO, Lincoln, NE, USA)
2. Prototype small-scale noodle-making equipment (Fig. 7)
3. TA-XT2 Texture Analyzer with special probes
4. Computer (programmed for data collection) or chart recorder
5. Water bath to maintain the temperature (at approximately 30°C) of the distilled water used for the test
6. Precision balance (readability of 0.01 g)

Fig. 7. Prototype small-scale noodle-making equipment developed by BRI Australia and CSIRO.

Procedure
The procedures of making WSN or YAN differ only in the formula of noodle dough.

Mixing
For WSN, 34.9% water absorption is used, the water contains 20% sodium chloride, while for YAN, a slightly smaller amount of water is used (32% absorption) as 10% sodium carbonate solution. (Amount of water is adjusted by the moisture content of the flour, determined by NIR). The mixing is carried out on a 10 g Mixograph, using 5 min mixing time at 66 rpm.

Sheeting
The sheeting step is the most critical phase of noodle making. The noodle piece is passed through the rolls step by step reducing the gaps between the rolls.
The process starts with a 3.8 mm gap. The steps of sheeting:
1. Pass the block through and retrieve it
2. Fold in half and pass through again, folded end first
3. Fold into thirds, (untidiest end on the inside) and pass through with the cut end up
4. Pass through again without folding
5. Place sheet in a sealed plastic bag and rest for 30 min exactly at room temperature (RT)
The same procedure is then repeated with 3.0, 2.0, and 1.2 mm roll gaps.
After removing the uneven edges the raw noodle is ready for texture analysis. If texture analysis is planned to be done on the cooked noodle, the noodle has to be cut a with hand-driven Otake machine and rested in a plastic bag for 2 h at 25°C before cooking.

Cooking
Both WSN and YAN are cooked the same way:
1. Boil 500 mL water in a small stainless steel saucepan on a gas stove.
2. Weigh 7.0 +/- 0.5 g of noodles and add to the rapidly boiling water
3. Stir briefly to loosen noodles and prevent them sticking to each other
4. Cook 6 min
5. Drain noodles and transfer to iced water for 2 min
6. Wash under a gentle stream of tap water for 1 min
7. Transfer to a bowl of water and wait for 2 min

Texture analysis
Special procedures developed for correlating to the mouth feel of noodle eating (Ross *et al* 1999) are used to evaluate noodle texture. The puncture test is used for raw noodle sheets and the compression and cutting tests for

cooked noodles. Settings for these tests measurements on the TA-XT2 Texture Analyzer are:

Test Conditions	Puncture	Compression	Cut
Test Mode and Option			
Measure Force in	Compression	Compression	Compression
Parameters			
Pre-Test Speed	10mm/s	10mm/s	10mm/s
Test Speed	3mm/s	1.6mm/s	0.8mm/s
Post Test Speed	10mm/s	10mm/s	10mm/s
Rupture Test Distance	1mm	1.00%	1.00%
Distance	20.0mm	50.00%	80.00%
Force	3N	0.098N	0.098N
Time	0.01 s	0.01 s	4.00 s
Count	2	2	2
Load Cell	5-0.1	5 -0.1	5 - 0.1
Temperature			
Trigger	0.01N	0.01N	0.01N
Units			
Force	N	N	N
Distance	Mm	%Strain	%Strain

Incorporation Techniques

The alteration of protein composition of flour by supplementation with different proteins is one of the most important *in vitro* methods available to determine their contributions to functional properties. The effect of added gliadin on dough mixing properties (MacRitchie, 1987) is the production of weaker and less stable doughs, as shown by decreases in mixing time and maximum resistance and an increase in resistance breakdown (Hussain and Lukow, 1997; Uthayakumaran *et al* 1999). Dough samples prepared in this way exhibit increased extensibility and poorer baking performance. In every case, the addition of lower molecular weight proteins effectively reduces the average molecular weight of the protein in the composite flour.

The effect of adding monomeric glutenin subunits to a "base" flour is also to effectively reduce the average molecular weight of the protein in the composite flour, because the relative amounts of polymer and monomer are shifted towards the monomer. In such an *in vitro* experiment, the added monomeric proteins would be expected to display properties characteristic of *in vivo* conditions only if they could form some part of the extended disulfide-linked gluten structure. Thus, meaningful estimates of the effects of added glutenin subunits on dough properties could be made only if they could be chemically incorporated into the glutenin polymer, analogous to their normal *in vivo* distribution.

Studies of the effects of a range of reductants and oxidants on the functionality of gluten proteins during dough mixing showed that it was possible to effectively destroy dough functionality with a reductant and then recover the functionality by subsequent oxidation (Békés *et al* 1994b). Careful selection of the oxidant, its concentration and reaction conditions allowed essentially complete recovery of the original dough mixing properties. This reduction/oxidation procedure has been used to incorporate a wide range of partially purified fractions or individual purified glutenin subunits into the polymeric glutenin phase (Békés *et al* 1994a, 1994c, Sapirstein and Fu, 1996, Veraverbeke *et al* 1999). The effects of the reduction/oxidation procedure on the mixing properties of mixtures of flour and added gliadin (Murray *et al* 1998) indicate that the effects of gliadin proteins on mixing properties are not altered by the reversible reduction/oxidation procedure. Thus, the presence of intra-molecular disulfide bonds in gliadin does not seem to interfere with the reduction/oxidation of glutenin.

The application of the reduction/oxidation procedure to dough preparations produced for extension measurement required different conditions to those for mixing studies, presumably because of continuing slow oxidation of dough components during the long relaxation required before stretching (Fig. 8).

The differences may offer further insight into the nature of the changes taking place between mixing and extension testing. Both the formulation and the reduction/oxidation conditions have to be modified if the resulting doughs are to be baked, because of the toxicity of dithiothreitol to yeast. Nevertheless, it has been possible to develop protocols in which functionality is maintained (Uthayakumaran *et al* 2000a,b).

The basic procedures for incorporating these glutenin subunits for mixing, extension and baking studies are described.

Mixing

All formulations are mixed in a 2 g Mixograph™ (TMCO, Lincoln, NE, USA). The blends (flour and the protein fraction), 450 μL of reducing agent (2 mg/ mL DTT solution - made in de-ionized water) and de-ionized water (calculated according to the AACC Method 54-40) are mixed for 30 s. The mixture is allowed to react for 4 min without further mixing. Towards the end of the 4th min the partially reduced dough is treated with 250 μL of oxidant (5 mg/ mL KIO_3 solution - made in de-ionized water), mixed for 30 s, rested for 5 min for the oxidizing reaction to take place and further mixed for 10 min.

Evaluation of the mixing curve after incorporation experiments requires a special treatment, cutting out the signal during the resting periods. The MD6 software has a built-in procedure for this.

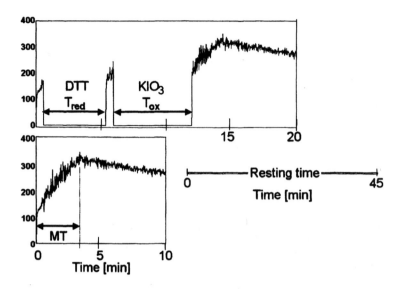

Fig. 8. Variables manipulated during incorporation of glutenin subunits. The top graph shows the mixing curve, interrupted for reduction and oxidation, and the bottom graph shows the reassembled mixing curve. The key variables are: time for reduction (T_{red}), reductant concentration (DTT), time for oxidation (T_{ox}) and oxidant concentration (KIO_3). In the case of incorporation for extension measurement, an additional parameter, the resting time can be altered to get optimal results.

Two different samples can be used for controls: flours treated the same way as above but without the addition of any glutenin subunits and flours mixed with water but following the same rest time protocol. The differences in the mixing properties of these two control samples demonstrate how efficient the reduction/oxidation conditions are in reversibly opening and closing the polymeric structure of glutenin. If more than 5% difference is observed between the mixing times in the two control samples, conditions have to be altered slightly. The paper of Békés et al (1994b) provides details how this alteration should be carried out.

Extension Testing
The reduction/ oxidation procedure developed by Uthayakumaran et al (2000a) can be used for incorporating different protein fractions into the flour for extension studies. Dough is prepared by mixing the flour, the protein to be incorporated, 450 µL of DTT solution (0.2 mg/ mL) plus the water required in a 2 g Mixograph. The dough is mixed for 30 s and then allowed to rest for one min. In the last few seconds of the resting period,

194

250 μL of oxidant solution, KIO$_3$ (5 mg/ mL) is added and mixing resumed for 30 s before the dough is allowed to rest for a further 5 min. The dough is then mixed to 70% of the peak dough development time (including the initial 2 × 30 s mixes). The total quantity of water to be used is calculated, using the protein and moisture contents of ingredients (Method 54-40, AACC, 1988). Dough samples (1.7g / test) are molded into cylinders approximately 6 mm diameter with a prototype molder, mounted on a sample carrier and rested at 30°C and >90% RH for 45 min before extension testing (see micro extension testing). The evaluation of the extension curve and the determination of parameters are identical to those described previously.

Micro-Baking

The reduction/ oxidation procedure developed by Uthayakumaran *et al* (2000a) can be used for incorporating different protein fractions into the flour for micro-baking studies. The total quantity of water to be used to prepare dough for microbaking is calculated using the protein and moisture contents of ingredients (AACC Method 54-40). The recipe used for baking is as given above. Flour and subunits to be incorporated, 450 μL of DTT (2 mg/ mL) solution plus the water (required water minus liquid contributed from DTT solution, KIO$_3$ solution and yeast solution) are weighed into the mixing bowl. The mixture is mixed for 30 s and allowed to rest for one min. In the last few seconds of the resting period, 250 μL of oxidant solution, KIO$_3$ (2.5 mg/ mL) is added with the yeast solution (which contains yeast, salt solution and improver. Mixing is resumed for 30 s and the dough allowed to rest for further 5 min. The dough is then mixed to the peak dough development time (including the initial 2 × 30 s mixes). Loaves are prepared from 2.4 g of the resulting dough as described previously. Loaf height is measured with vernier calipers.

Acknowledgement

The authors wish to acknowledge K. Adams (Cereal Research Centre, Agriculture and Agri-Food Canada), B. Butow (CSIRO Plant Industry, Canberra, Australia) and R. Haraszi (TUB, Budapest, Hungary) for their valuable contribution to produce this chapter. We also wish to thank P.W. Gras for his comments and help.

References

American Association of Cereal Chemists. 2000. Approved method of the AACC. Method 10-10, Method 44-15A, Method 54-21, Method 54-40, 74-09 The Association: St. Paul, MN.

Békés F, Anderson O, Gras PW, Gupta RB, Tam A, Wrigley C, Appels R. 1994a. The contribution to mixing properties of 1D glutenin subunits expressed in a bacterial system. Pages 97-104 in Improvement of Cereal Quality by Genetic Engineering. R. Henry and J.A. Ronalds eds. Kluger, London.

Békés F, Gras P.W., Gupta R.B. 1994b. Mixing properties as a measure of reversible reduction/oxidation of doughs. Cereal Chem. 71:44-50.

Békés F, Gras P.W., Gupta R.B., Hickman D.R., Tatham A.S. 1994c. Effects of 1Bx20 HMW glutenin on mixing properties. J. Cereal Sci. 19:3-7.

Békés, F., and Gras, P.W. 1999. In vitro studies on gluten protein functionality. Cereal Foods World, 44:580-586.

Békés, F., and Gras, P.W. 2000. Small-scale dough testing as a breeding and research tool. Chem. Australia 67:33-36.

Butow, B.J., Gras, P.W., Murray, D., Quail, K., Hogan, W., and Békés, F. 2002. Sponge and dough method by micro-baking. Proc. 51st RACI Cereal Chemistry Conference, 2001, Bronte.

Gras, P.W., Anderssen, R.S., Keentok, M., Békés, F., and Appels, R. 2001 Gluten protein functionality in wheat flour processing. Aust. J. Agric. Res. 52:1313-1323.

Gras, P.W., and Békés, F. 1996. Small-scale testing: The development of instrumentation and application as a research tool. Pages 506-510 in Proceedings of the 6th International Gluten Workshop. C.W. Wrigley ed. Royal Australian Chemical Institute, North Melbourne, Australia.

Gras, P.W., Hibberd, G.E., and Walker, C.E. 1990. Electronic sensing and interpretation of dough properties using a 35-g Mixograph. Cereal Foods World 35:568-571.

Gras, P.W., Varga, J., Rath, C., Tomoskozi, S., Fodor, D., Nanasi J., Salgo, A., and Békés, F. 2000. Screening for improved water absorption and mixing properties using 4g of flour: a new small-scale Farinograph-type mixer. Pages 174-179 in Proc. 11th Cereal and Bread Congress, Cereals, Health and Life, Surfers Paradise. M Wooton, IL Batey, and CW Wrigley eds. RACI, 2001, Melbourne

Haraszi, R., Tomoskozi, S., Gras, P.W., and Békés, F. 2001. A research tool to study dough properties: the small-scale Z-arm mixer. In Proc. 51st RACI Cereal Chemistry Conference, Bronte. M Wooton, IL Batey, and CW Wrigley eds. RACI, Melbourne

Hussain, A., and Lukow, O.M. 1997. Influence of gliadin-rich subfractions of glenlea wheat on the mixing characteristics of wheat flour. Cereal Chem. 74:791-799.

Ingelin, M.E., and Lukow, O.M. 1999. Mixograph absorption determination by response surface methodology. Cereal Chem. 76:9-15.

Kieffer, R., Garnreiter, F., and Belitz, H.D. 1981. Beurteilung von Teigeigenschaften durch Zugversuche im Mikroma βstab. Z. Lebensm. Forsch. 172:193-194.

Kilborn, R.H., and Tipples, K.H. 1981. Canadian test baking procedures. I. GRL Remix Method and variation. Cereal Foods World 26:624-627.

Konik, C. M., Miskelly D.M., and Gras, P.W. 1992. Contribution of starch and non-starch parameters to the eating quality of Japanese white salted noodles. J. Sci. Food Agric. 58:403-406.

MacRitchie, F. 1987. Evaluation of contributions from wheat protein fractions to dough mixing and bread making. J. Cereal Sci. 6:259-268.

Mann, G. 2001. New Crop Phbenomenon in wheat and the mechanisms involved. PhD. Thesis. University of Reading.

Miskelly, D.M., and Moss, H.J. 1985. Flour quality requirements for Chinese noodle manufacture. J. Cereal Sci. 3:379-387.

Murray, D.J., Békés, F., Gras, P.W., Copeland, L.A., Savage, A.W.J., and Tatham, A.S. 1998. Hydrogen bonding and the structure/function relationships of wheat flour gliadins. Pages 12-16 in 'Cereals 98 - Proc. 48th RACI Cereal Chemistry Conference'. L O'Brien, AB Blakeney, AS Ross and CW Wrigley eds. RACI, North Melbourne.

Pon, C.R., Lukow, O.M., and Buckley, D.J. 1989. A multichannel, computer-based system for analyzing dough rheology. J. Texture Stud. 19:343-360.

Preston, K.R., Kilborn, R.H., and Black, H.C. 1982. The GRL pilot mill. II. Physical dough and baking properties of flour streams milled from Canadian Red Spring Wheats. Can. Inst. Food Sci. Tec. J. 15:29-36.

Preston, K.R., and Hoseney, R.C. 1991. Applications of the Extensigraph. Pages 13-19 in The Extensigraph Handbook. F Rasper and KR Preston eds. American Association of Cereal Chemists. St. Paul, MN.

Quail, K., Jin, Y., Rema, G., Yun, H., Caldwell, R., Békés, F., and Partridge, S.J. 1999. Effect of starch and gluten addition on noodle texture. Pages 78-81 in Cereals '99, Proc. 49th Aust. Cereal Chem. Conf. JF Panozzo, M Ratcliffe, M Wootton, and CW Wrigley eds. RACI, North Melbourne, Australia.

Rath, C.R., Gras, P.W., Wrigley, C.W., and Walker, C.E. 1990. Evaluation of dough properties from two grams of flour using the Mixograph principle. Cereal Foods World 35:572-574.

Rath, C.R., Gras, P.W., Zhen, Z., Appels, R., Békés, F., and Wrigley, C.W. 1994. A prototype extension tester for two-gram dough samples. Pages 122-126 in Proceedings of the Australian Cereal Chemistry Conference, 44th. JF Panozzo and PG Downie eds. Royal Australian Chemical Institute, North Melbourne, Australia.

Ross, A.S., Quail, K.J., Crosbie, G.B. 1996. An insight into structural features leading to desirable alkaline noodle texture. Pages 115-119 in Proc. 46th RACI Conference. CW Wrigley ed. RACI, Melbourne

Ross, A.S., To, S., Chiu, P.C., Quail, K.J. 1999. Instrumental evaluation of white salted noodle texture. Pages 199-204 in Proc. 48th RACI Conference. AW Tarr, AS Ross and CW Wrigley eds. RACI, Melbourne

Sapirstein, H.D., and Fu, B.X. 1996. Characterization of an extra-strong wheat: Functionality of 1) gliadin- and glutenin-rich fractions, 2) total HMW and LMW subunits of glutenin assessed by reduction-reoxidation. Pages 302-306 in Proc. 6th International Gluten Workshop. CW Wrigley ed. Royal Australian Chemical Institute, North Melbourne, Australia.

Smewing, J., 1995. The measurement of dough and gluten extensibility using the SMS/Kieffer rig and the TA.TX2 texture analyser handbook. SMS Ltd.

Suchy, J., Lukow, O.M., and Ingelin, M.E. 2000. Dough microextensibility method using a 2-g mixograph and texture analyzer. Cereal Chem. 77:39-43.

Uthayakumaran, S., Gras, P.W., Stoddard, F., and Békés, F. 1999. Effects of varying protein content and glutenin-to-gliadin ratio on the functional properties of wheat dough. Cereal Chem. 76:389-394.

Uthayakumaran, S., Stoddard, F.L., Gras, P.W., and Békés, F. 2000a. Optimized methods for incorporating glutenin subunits into wheat dough for extension and baking studies. Cereal Chem. 77:731-736.

Uthayakumaran, S., Stoddard, F.L., Gras, P.W., and Békés, F. 2000b. Effects of incorporated glutenins on functional properties of wheat dough. Cereal Chem. 77:737-743.

Veraverbeke, W.S., Verbruggen, I.M., and Delcour, J.A. 1999. Effects of increased HMW-GS content of flour on dough mixing behavior and breadmaking. J. Agric. Food Chem. 46:4830-4835.

Voisey, P.W., Miller, H., and Kloek, M. 1969. The Ottawa electronic recording dough mixer. VI. Differences between mixing bowls. Cereal Chem. 46:196-202.

Yun, S.H., Rema G. and Quail, K. J. 1998. Wheat starch fractionation.. Pages 365-368 in Proc. 47th RACI Conference. AW Tarr, AS Ross, and CW Wrigley eds. RACI, Melbourne

Zhao, X.C., Batey, I.L., Sharp, P.J., Crosbie, G., Barclay, I., Wilson R., Morell, M.K. and Appels, R. 1998. A Single Genetic Locus Associated with Starch Granule Properties and Noodle Quality in Wheat. J. Cereal Sci. 27:7-13.